Exclusively endorsed and approved by AQA

Gerry Blake • Jo Locke

Series Editor: Lawrie Ryan

GCSE Additional Applied Science

Published in 2006 by:
Nelson Thornes Ltd
Delta Place
27 Bath Road
CHELTENHAM
GL53 7TH
United Kingdom

07 08 09 10 / 10 9 8 7 6 5 4 3 2

A catalogue record for this book is available from the British Library

ISBN 978 0 7487 9655 7

Cover photographs: wheat field by Corel 555 (NT); fingerprints by Corel 565 (NT); snowboarder by Digital Vision XA (NT).

Cover bubble illustration by Andy Parker

Illustrations by Oxford Designers and Illustrators and Kevin Jones Associates

Page make-up by Design Practitioners Ltd

Printed and bound in Slovenia by Koratan – Ljubljana Ltd

GCSE Additional Applied Science

Contents

GCSE Additional Applied Science

Welcome to AQA Science!

LEARNING OBJECTIVES

By the end of the lesson you should be able to answer the questions posed in the learning objectives; if you can't, review the content until it's clear.

GET IT RIGHT!

Avoid common mistakes and gain marks by sticking to this advice.

PRACTICAL

You'll find lots of practicals to help you with your learning.

DID YOU KNOW?

Curious examples of scientific points that are out of the ordinary, but true…

COURSEWORK HINTS

This feature provides useful hints and reminders on completing your coursework.

COURSEWORK TASKS

This feature appears on the coursework pages that are designed to guide you through your coursework investigation.

KEY POINTS

If you remember nothing else, remember the key points! Learning these in each lesson is a good start. They can be used in your revision and help you summarise your knowledge.

AQA GCSE Additional Applied Science is the new work-related qualification that puts science to work in an everyday context.

This textbook covers all three units of the specification and emphasises science in the world of work. It includes both scientific principles and laboratory procedures. Advice is given to help you produce quality coursework. In-text and summary questions allow you to check your understanding and progress. The book is broken down into five chapters with each idea set out in a double-page spread, starting with the 'Learning objectives'.

Your teachers will decide on the most appropriate practical work and coursework tasks for you to carry out. They will also assess your coursework **portfolio** for Units 1 and 3 of the specification. This work accounts for 60% of your GCSE marks. Your portfolio must be kept secure.

The three units of this course (as in the specification) are as follows:

Unit 1, 'Science in the workplace', carries 20% of the total marks. In this unit you write two reports, investigating:

- How science is used (Chapter 1) and
- Working safely (Chapter 2)

Unit 1 is entirely assessed on the evidence contained in your portfolio of work.

Unit 2, 'Science at work', carries 40% of the marks and is assessed by a 1 hour written examination paper, based on three topics:

- Food science (Chapter 3) – the work of food scientists and how their work encourages us to have a healthy diet.
- Forensic science (Chapter 4) – how scientists use forensic techniques to help solve crimes.
- Sports science (Chapter 5) – how sports scientists help with diet and fitness, and with the design of sports equipment and clothing used for sport.

Unit 3, 'Using scientific skills', carries 40% of the marks.

In this unit you produce a report of *one* practical investigation set in a vocational context covering *either* food science *or* forensic science *or* sports science.

Unit 3 is entirely assessed on the evidence contained in your portfolio of work.

You can find the details of each investigation in the following sections:

- Food science coursework (Chapter 3, pages 58–67).
- Forensic science coursework (Chapter 4, pages 100–11).
- Sports science coursework (Chapter 5, pages 146–55).

SUMMARY QUESTIONS

Did you understand everything? Get these questions right, and you can be sure you did. Get them wrong, and you might want to take another look.

1.1 Using science

LEARNING OBJECTIVES

1. What is involved in this coursework investigation task?
2. How do I build up a good portfolio?

People often think of scientists wearing white lab coats. However, we use scientific skills in so many jobs: vets, nurses, engineers, pharmacists, photographers, chefs, beauticians and gardeners, to name but a few.

Not all scientists wear lab coats

A range of scientific jobs

Science is big business in the UK.

a) Name a big manufacturing firm near where you live.
b) Name a large employer in your area that uses science and provides a service to the public.
c) How many people in your class know friends or family who work in these two workplaces?

As Applied Science students you will write a portfolio report *(Unit 1 – Task A)* on 'How Science is Used in the Workplace'. You will do research using books and computers. You will also interview people and write letters to companies to find out information.

d) Discuss the best methods to get different types of information for a project. Then draw a table showing different methods of collecting information and when you have used them.

Method of collecting information	Where I have used it before
Surveys	Geography project on where people live

Making an excellent portfolio

The most important thing to remember is that other people are going to read your report. They will make decisions based on the information you give them. Imagine inventing the best ever engine but trying to sell it in a really ugly and unstylish car. It wouldn't sell. To be successful, make sure your ideas are both good and well presented.

The **assessment evidence grid** (mark scheme) for this report follows:

1A *You should be able to*:	2A *You should be able to*:	3A *You should be able to*:
• produce a simple study on a range of organisations that use science • state the products made or services provided • identify the jobs of those employed	• identify organisations as local, national or international • describe their location • describe the products made or services provided • describe the jobs and qualifications of the employees and how they use science • describe the types of skills scientists need in addition to their qualifications, and a range of careers that are available in science	• produce a simple study of ***one*** particular organisation • explain its location • describe the products made or services provided and explain their importance to society • give a detailed account of the skills and qualifications needed by scientists who work there • describe the effect on the local environment of the organisation
1–3 marks	4–8 marks	9–11 marks

'How Science is Used' coursework checklist

1. Identify one local, one national and one international organisation that use science.
2. What does each organisation do? Describe the products they make and/or the services they provide.
3. Where are the organisations located and why are they located there?
4. Explain their importance to society and their effect on the local community.
5. How many people does each organisation employ?
6. Put the employees of each organisation into one of three classes: major, significant and small users of science.
7. Identify the job titles and qualifications of the people who perform scientific tasks.
8. Describe the types of scientific activity they carry out on a daily basis.
9. What skills do the scientists need in addition to their qualifications?
10. As an appendix to your report, write about the careers that are available in science-related areas and their importance.

Students must show that they have studied one organisation in detail to obtain a stage 3A mark.

COURSEWORK TASKS

1. Which friends or family members work in an organisation that uses science? (You could interview them.)
2. Which three organisations could you use in the project?
3. Using the checklist above as a guide, write a questionnaire to help you get the information you need from the three organisations.

Back to the drawing board?

GET IT RIGHT!

- Use a mind map to help plan your work.

- Divide up your work into easy to understand sections.
- Include copies of letters, questionnaires, and replies from organisations, etc., as appendices to your report. Write down all your sources of information at the end.
- Write the contents page and put the page numbers on last, after you have finished your project.

1.2 Reporting to impress

LEARNING OBJECTIVE

1 How can I get information for my portfolio?

How clearly you present your information does affect your coursework marks. It is also important to use a wide range of sources and information. See the 'assessment evidence grid' on the previous page.

Interviewing – Asking a scientist questions about their work is a very good way of getting information for your portfolio. Prepare your questions and include them, as well as the replies you got, in your portfolio.

E-mails and letters – If you e-mail or send letters to a company to ask questions about their work, make sure you include your e-mail or letter in your portfolio.

COURSEWORK HINTS

Also ask about other skills and qualities that are needed, e.g.

Skills	Qualities
Interpersonal	Patience
Communication	Reliability
Numeracy	Initiative
Organisational	Flexibility
ICT	Accuracy
Imaginative	Caring

Ask a librarian – Ask a librarian to help you search for the right information.

Work experience – There is always some science that is going on in every business. At the start of your work experience, make sure you find out all about the health and safety issues. You should include this information in your portfolio.

Surveys – Surveys are very effective to get lots of information from lots of people quickly. You should show your teacher the questions that you are going to ask before you start. By asking multiple choice questions, you can analyse the answers you get if your sample is large enough.

Websites – Websites often show lots of information about companies. It is important that you don't just download and print off loads of pages. The projects you do should show your ideas, not someone else's.

1C *You should be able to*:	2C *You should be able to*:	3C *You should be able to*:
• use a limited range of sources and information, to present findings in your portfolio 1 mark	• use a range of sources and information, to present findings clearly in your portfolio 2 marks	• identify and use a wide range of sources and information, to present findings clearly and logically throughout your portfolio 3 marks

1.3 Location, location, location

LEARNING OBJECTIVES

1. How does location affect a business?
2. How can I include information about location in my Unit 1 portfolio?

The location of a business has a huge impact on how that business performs.

A town centre location would allow businesses easy access to potential customers. Likewise, an out of town shopping complex would be easy for people to visit and park their cars. This is not useful for some businesses, as they may need excellent road and rail links for staff and to transport their products.

Research or business parks are useful for companies who need lots of parking for their staff, but customers may find it difficult to get there. Internet-based companies and research companies tend to be in these parks.

a) Find out about a local business park. What businesses are there, and what do they do?

An out of town shopping complex

Out of town shopping malls are very popular shopping and leisure destinations.

b) What sorts of businesses would be successful at an out of town shopping centre? What sorts of businesses would be unsuccessful?

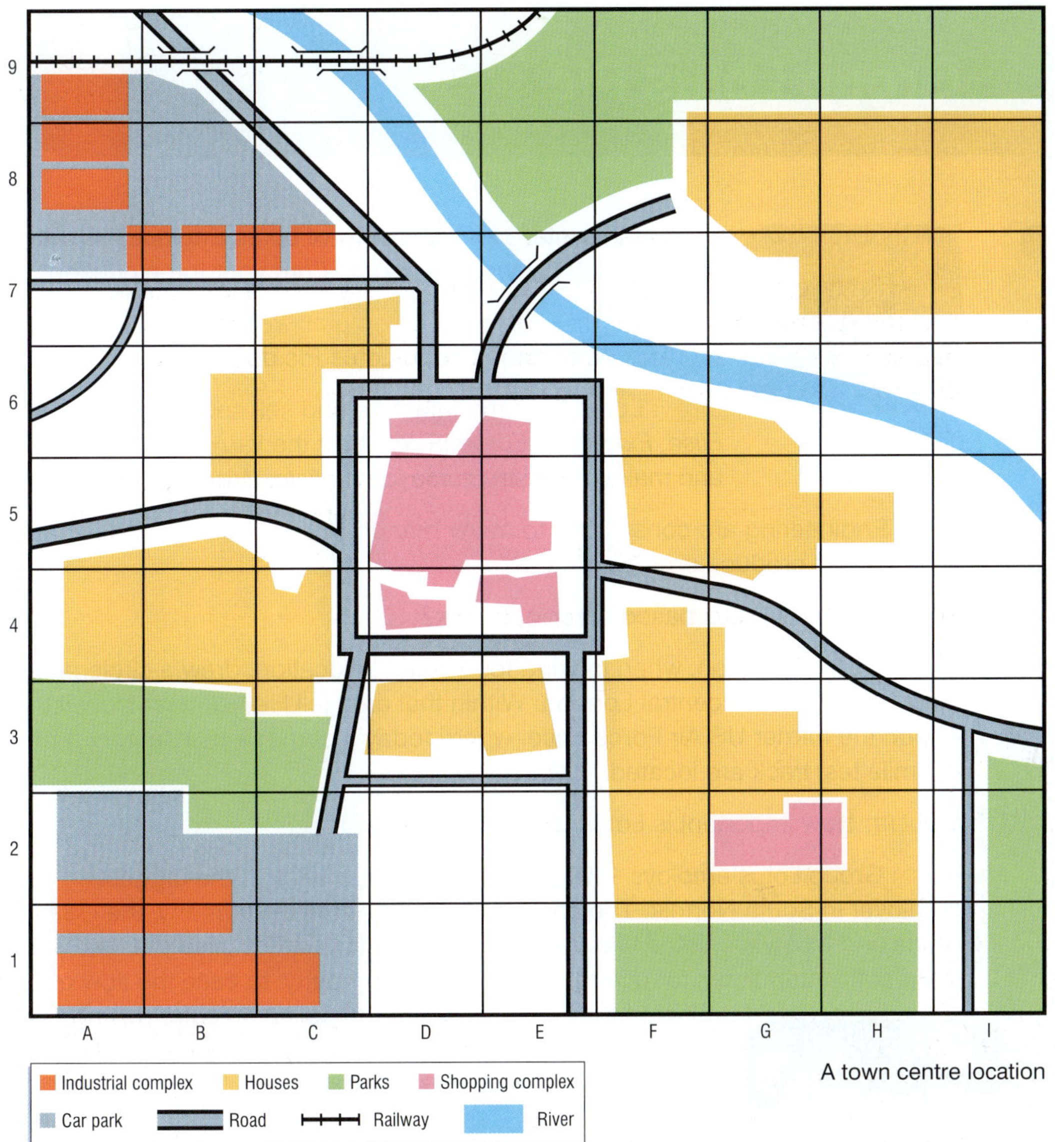

A town centre location

c) Copy the table.
   (i) Match the businesses to a suitable location from the list:
   town centre, shopping centre (out of town), business park, local shops.
   (ii) Where would you put the businesses on the map? Use the map references to help.

Business	(i) Location	(ii) Map reference
Optician		
Chemist		
Chemical factory		
Gymnasium		
Doctors' surgery		
New university		
Car manufacturer		
Greengrocer		
Boat builder		
Car mechanic		

COURSEWORK TASKS

Look at the information you need for your project on page 3.

1 What three organisations are you researching for your project? (*Remember: one local, one national and one international.*)

2 Describe and explain the location of these three scientific workplaces.

1.4 Group Lotus Plc

LEARNING OBJECTIVE

1 Support for your 'How Science is Used' coursework task – what response can you expect from an interview or questionnaire?

COURSEWORK HINTS

For your portfolio choose workplaces:

1 near to where you live
2 where you know the name of a contact
3 where you can visit.

Lotus logo

Look at this interview with Laura of Lotus Engineering. Plan your own interviews with people from your three chosen workplaces.

STUDENT: What does Group Lotus Plc do?

LAURA: Lotus Cars manufacture and sell sports cars, the *Elise*, *Exige* and *EuropaS*. We built the *Vauxhall Speedster* and make crash structures for the *Aston Martin V12 Vanquish*.

Lotus Engineering are consultants to many manufacturers, helping to design, develop and produce prototype cars.

STUDENT: Why is Lotus based in rural Norfolk?

LAURA: Colin Chapman, when looking for a suitable location, drew a circle on a map 100 miles from central London. Within that area fell Hethel, near Norwich. It is on the former US Air Force base, where today's purpose-built factory and 2.5-mile test track are located.

STUDENT: How important is Lotus to the local community?

LAURA: Group Lotus employs 1100 people at Hethel, making it the biggest employer in South Norfolk. The company also has staff in Germany, the USA, China and Malaysia. Lotus tries to be a good neighbour. For example, we stagger the start times for different areas of the company, to ease the flow of traffic. Lotus also works with the police on safer driving initiatives. This includes providing a police-branded *Lotus Elise* for summer 'car cruise' events.

Colin Chapman moved Lotus to Hethel, near Norwich in Norfolk, in 1966

STUDENT: Do all employees use science?

LAURA: All employees of Lotus are involved with science at some level. Everybody understands the scientific concepts behind the cars.

Major users of science	Significant users of science	Small users of science
Engineers	Factory workers	Directors
Technicians	Designers	Sales & Marketing Dept.
		PR team
		Administrative roles

STUDENT: What are the qualifications of the people who do scientific tasks?

LAURA: Not everybody working at Lotus has a degree in engineering, but many receive on-the-job training and study for qualifications throughout their time here. Engineers carry out various scientific tasks, depending on their specialism. We have designers making 3D digital images, clay modellers, and business graduates working with our technical specialists.

STUDENT: Can you describe some of the other scientific activities that Lotus carries out on a daily basis?

LAURA: Lotus engineers carry out many automotive activities, from crash-testing new components to improving engine management systems. From the chassis and suspension to the engine and traction control – every aspect of the car's make up – our engineers at Lotus can design from scratch. We use the latest computer systems when designing 3D models. This is not just for aesthetic perfection, but also to improve the aerodynamics and performance.

Scientific developments also mean our technicians now bond the aluminium chassis together using adhesive. This produces a stronger join than by welding. Similarly, we discovered that water-based paints create the best finish. They also do less damage to the environment.

STUDENT: Apart from their scientific qualifications, what skills do your employees need?

LAURA: Lotus 'scientists' need vision and adaptability to 'change the rules' and look at new solutions to engineering projects. We are an innovative company, always aiming for top-quality products.

STUDENT: Laura, thank you so much for your help today.

Precise measurements – engine design

A *Lotus Elise* being tested in an 'anechoic chamber', where no sound can echo off the walls of the room

Design in clay in the Lotus design styling studio!

COURSEWORK TASKS

1 Remember you are studying one local, one national and one international organisation. Who are your contacts in your chosen organisations?

2 Plan your interview questions and arrange your meeting.

(Remember: you must study one organisation in more detail to obtain a stage 3A mark – see the table on page 3.)

DID YOU KNOW?

Proud years for Lotus

Emerson Fittipaldi, 1972, one of the six World Drivers' Champions for Team Lotus

Chris Boardman on his 1992 Olympic winning Lotus bicycle

1.5 Tapping into your talents

Where do you want to take your future? Now is the time to think about where you want your next steps to take you. Many opportunities are available to you, so that you can become suitably qualified and skilled to get a career that you'll be happy with.

You may not have any idea as to what your next steps are, let alone choice of career. If that's the case, there is an information booklet written especially for Additional Applied Science students like you. Ask your careers office for the *Directions* booklet produced by SEMTA. It is subtitled, *Opportunities after GCSEs*, or find it with a Google search.

(See also *Connexions Direct* for information and advice about making decisions and choices in your life.)

Everyday life relies on science and technology. As we make our phone calls, travel to work, school or on holiday, do our jobs or enjoy entertainment, we connect to invention, manufacture and science. Science is not, of course, just a textbook subject, but a way of exploring everyday life and the world around us, and these are especially exciting times for science. Increasingly, women choose scientific careers.

A display showing up to date bus timetable information

Global positioning system (GPS) satellites determine the location of buses. Real-time data, an appliance of science, keep travellers up to date with travel information.

Miah is a dispensing technician, who works in a chemist's shop.

STUDENT: What do you do?

MIAH: Customers bring in their prescriptions and I prepare them exactly as specified. It is a responsible job as the well-being of our customers depends upon my accuracy. I also get involved in serving at the counter. Some days can be very busy and tiring.

STUDENT: Do customers ever ask for advice?

MIAH: Yes, counselling people is part of my job. I have to advise them whether or not they should seek medical advice.

STUDENT: How did you get into this work?

MIAH: I began work as an assistant in the chemist's. I was promoted to 'dispenser' after I qualified. I get on-the-job training with the company.

STUDENT: What skills and qualities are needed for this type of work?

MIAH: Accuracy and attention to detail are the most important, as mistakes mustn't happen. Also good communications skills are needed to interact with the customers. There are no minimum entry requirements for a job as a pharmacy technician, although employers and colleges expect some A–C GCSE grades.

A pharmacy technician explains a medicine to a customer

2.1 Working safely – the hidden costs if you don't

LEARNING OBJECTIVES

1 What are the hazards, risks, injuries and illnesses associated with the workplace?
2 What is involved in this coursework investigation task?

Power cables

Imagine that you live under these power cables. The high voltage could be ***hazardous*** (dangerous), as there is a slight ***risk*** (or chance of being harmed) if a pylon fell down in a storm.

There are hidden dangers in the workplace. More than 2500 people in the 16 to 24-year-old age group will be seriously injured at work this year, involving broken bones and serious burns.

a) How do we all take risks in life?
b) Why do you think young people face the highest risk of injury?

Workplace hazards

We cannot remove the hazards completely, but we can reduce the risks.

Think about the main reasons why people get injured:

- Slips and trips – 33% of all major injuries. The cost to employers is £512 million and the cost to the heath service is £133 million, per year. The human cost cannot be calculated.
- Unsafe lifting and carrying, causing back pain.
- Falling from a height – causing 70 deaths and 4000 serious injuries each year.
- Being struck by a moving object. Each year, 3500 people are killed on our roads and 40 000 are seriously injured.

More than 1 million people have days off work each year because of problems with muscles and joints.

Half a million people are ill because of work-related stress. Although pressure keeps us motivated, excess pressure causes illness.

In scientific workplaces, we have hazardous substances and dangerous equipment to deal with. There are also electrical, noise and manual handling problems. Scientific work can be dangerous.

c) Why do you think accidents among scientists are rare?
d) How could you avoid having a certain accident and so reduce the risk to yourself and to others? (Think about protective clothing, checking equipment and training.)

COURSEWORK HINTS

1 Talk to your school or college technicians about working safely.
2 Plan a workplace visit.
3 Write a questionnaire on working safety.
4 Try these websites:
www.hse.gov.uk
www.cleapss.org.uk
5 See the CLEAPSS CD-ROM.

The **assessment evidence grid** (mark scheme) for this report follows:

1B *You should be able to:*	**2B *You should be able to:***	**3B *You should be able to:***
• carry out research into working safely in the school or college laboratory, including hazards and risks and their assessment, first aid and fire prevention 1–4 marks	• carry out research into the issues of working safely in a workplace that uses science or scientific skills, including hazards and risks and their assessment, first aid and fire prevention 5–8 marks	• carry out research into the issues of working safely in a scientific workplace and compare these with the school or college laboratory, including hazards and risks and their assessment, first aid and fire prevention 9–11 marks

WORKING MORE INDEPENDENTLY →

In the school or college laboratory, you could start by identifying:

- Hazard warning signs.
- Biological, chemical and physical hazards, including radioactive substances, and their risks.
- Health and safety procedures.
- How we use **risk assessments**.

You can find out more about your chosen scientific workplace by identifying:

- Health and safety checks.
- Risk assessments for tasks carried out.
- What they do to prevent accidents from the hazards that exist.
- Emergency procedures followed if an accident does happen.

As Applied Science students you will write a report *(Unit 1 – Task B)* on:

- 'Working Safely in a Scientific Workplace', comparing this with
- the health and safety precautions in your own school or college.

Worker on a girder at a construction site

COURSEWORK TASKS

Ideas for your questionnaire about health and safety at your chosen scientific workplace

- What hazards have you identified at *[name of company]*?
- What have you done to control the risk of these hazards?
- What new safe working practices have you introduced, if any?
- Have these reduced accidents, absence rates and saved your company money?

Discuss with a partner what other questions you could ask for your chosen workplace. Produce your own questionnaire.

DID YOU KNOW?

Sarah was using an unguarded drilling machine. The sleeve of her jumper caught on the rotating drill, entangling her arm. Both bones in her lower arm were broken and she suffered extensive tissue and muscle injury. She spent 10 days in hospital and was off work for 3 months. She was unable to operate machinery for 8 months. Her manager was prosecuted.

The cost to the business was £45 000. Two other employees were made redundant to prevent the company going out of business.

2.2 Laboratory safety – who is it for?

LEARNING OBJECTIVES

1. What health and safety issues exist in science laboratories?
2. What hazard warning signs do we use?
3. Why are radioactive materials hazardous?

In 1911, Marie Curie won her second Nobel Prize for her work on radioactivity. She started the use of mobile X-ray units to help wounded soldiers in the First World War. Almost 100 years later, X-rays are common. Marie Curie did not know the dangers of radiation when she first started working with radioactive materials. Sadly, Marie died of leukaemia (cancer of the blood) due to radiation exposure.

Marie Curie

Today, doctors get the clearest medical images with positron emission tomography (PET) scans. Doctors inject very small amounts of radioactive materials into the patient. Gamma-ray cameras placed round the patient can then show exactly where diseases such as cancer are. Notice that the technicians work behind a protective screen, where they will be joined by the nurse.

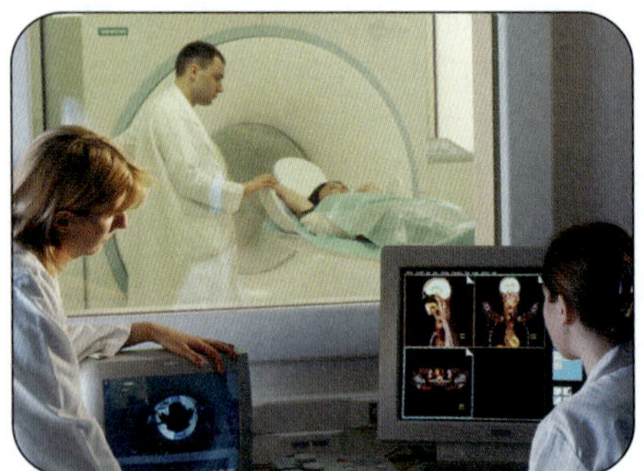
A PET scan machine

a) Why do you think they need a safety screen?

Imagine in 5 years time – you could be working in a laboratory preparing materials to use in hospitals!

You have probably seen an experiment with a Geiger counter.

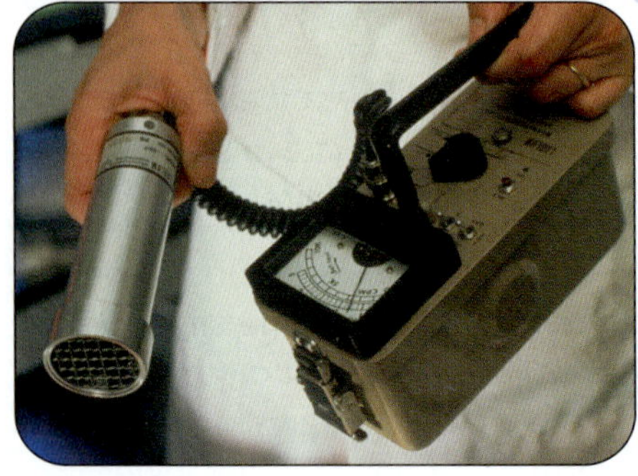
A Geiger counter

b) What safety precautions did your teacher take?
c) Look at the two photographs below. How do the suits protect the workers from radiation?

Space suit

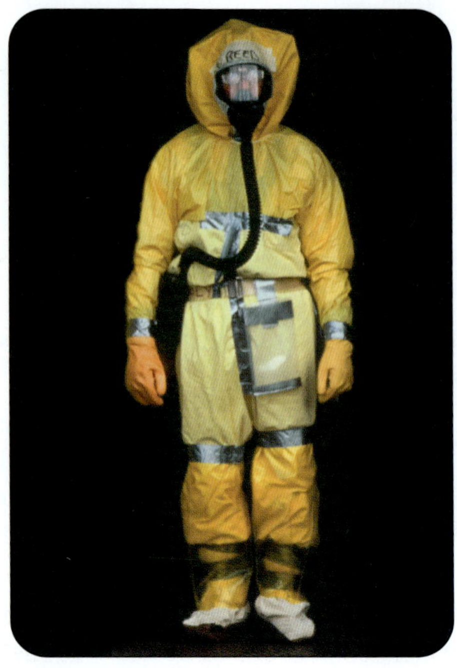
Hostile environment suit

Hazard symbols

- Yellow and black triangles ***warn of danger***, e.g. radioactivity.
- A red crossbar is something you ***must not do,*** e.g. no smoking.
- A blue circle is something you ***must do***, e.g. wear eye protection.
- A green background gives ***safety information***, e.g. first aid.
- ***Hazard labels*** are orange and black, e.g. **flammable**.

Hazard warning signs

You already know some risks of using poor techniques in experiments.

The school laboratory technicians have to be careful too, when they prepare chemicals. For example, sodium hydroxide is corrosive, causing severe burns. It is particularly dangerous to eyes. It gets hot when added to water.

Sodium hydroxide is an alkali. You sometimes use sodium hydroxide solution in science experiments.

Technicians prepare your solutions by dissolving solid sodium hydroxide in water.

d) When preparing solutions of sodium hydroxide, what safety precautions should the technician use to reduce the risk?
e) Some sodium hydroxide solution spills on the table. What do you do if you accidentally wipe some NaOH solution in your eye?

DID YOU KNOW?

High levels of ionising radiations can damage body cells, cause loss of hair and radiation sickness. They also cause cancer.

Some people work with radioactive substances. Not surprisingly, their levels of exposure to radiation are checked.

They wear 'film badges' that are sensitive to radioactivity.

COURSEWORK TASKS

1 What do these words mean?

a) inflammable b) biohazard c) toxic d) harmful

e) corrosive f) irritant g) explosive h) oxidising.

2 The last five warning labels do not match with the hazard. What is the correct order?

Biohazard	Harmful irritant	Corrosive	Toxic	Oxidising	Explosive

	A	B	C	D	E

3 Find out how your laboratories are protected. Consider mains electricity, gas, fumes and explosions.

4 Think about the safety rules in your school or college laboratory. What five rules do you think are the most important and why?

5 Find out about how unwanted or waste materials, including radioactive substances, are disposed of safely.

COURSEWORK HINTS

Organise a tour of the laboratories and school/college preparation rooms with your science technicians.

2.3 Risk assessments – thinking first

LEARNING OBJECTIVES

What are the following:

1. Hazards and risks?
2. Accident prevention?
3. Health and safety?
4. Risk assessments?

An unexpected hazard

Fork-lift truck driver

DID YOU KNOW?

There are 1.6 million injuries at work each year.

70% of these could be prevented if employers put control measures in place.

The rate of injuries in firms with fewer than 50 employees is over twice the rate of that in firms employing more than 1000 people.

Find out lots more at www.youngworker.co.uk

Some people like taking risks, e.g. skiing or bungee jumping.

Sometimes the hazards come with the job, like being a policeman.

Sometimes we are not aware there is a risk.

A ***hazard*** is anything that can cause harm. Examples include chemicals, electricity and gas, noise, careless behaviour, microorganisms, etc. The ***risk*** is the chance that someone will be harmed by the hazard.

a) Think of a fork-lift truck (see the photograph). When might it pose a risk?
b) Who might be harmed?
c) How could accidents with fork-lift trucks be limited?

250 people each year lose their lives at work in Britain. More than 150 000 are injured or badly hurt.

Safety slogans

In your school or college laboratory, minor burns to hands caused by touching hot tripods and test tubes are fairly common. The hazard (or danger) is minor, as the burns are small and heal quickly. The risk is high, but the consequences are not very serious.

d) What ***control measures*** (or safety precautions) would you use to reduce the number of minor burns in schools and colleges?
e) What is the ***emergency action*** (first aid) for a minor burn accident?

Electric shock from a faulty appliance could result in death. It is a major hazard. However, with properly designed and maintained equipment, the risk is insignificant.

Concentrated sulfuric acid is corrosive. It causes severe burns and can permanently damage eyes. It is a major hazard. If it is left out in a laboratory, the risk is significant.

f) What control measures can we use to reduce the risk when using sulfuric acid? (See the CLEAPSS CD-ROM Student Safety Sheet 22.)

We make risk assessments to ensure that we control or get rid of risks. If you think it's just a form-filling exercise you've missed the point! In a risk assessment we:

1 Identify ***hazards***, i.e. things that could cause harm – materials, procedures and equipment. (See question 1 below.)

2 Work out the ***risk*** – how likely is it that harm could occur and how serious could the consequences be?

3 Put ***control measures*** (safety precautions) in place to avoid or reduce the risk as far as possible.

4 Decide on the ***emergency action*** (first aid) to take if the controls fail and there is an accident.

The table below shows how to set out a risk assessment:

Risk Assessment Form Name of student: ____________ Date: ____________

Task: ____

Hazards	Risks	Control measures	Emergency action
Think about the material, procedure or equipment. What makes it dangerous and what could go wrong?	(high / moderate / low) Probability of harm and seriousness of consequence	Safety precautions	First aid

You will find the terms hazard, risk and risk assessment in health and safety legislation, like the Control Of Substances Hazardous to Health (COSHH) regulations.

The job of the Health and Safety Executive (HSE) is to see that risks to people's health and safety from work activities are properly controlled.

See www.hse.gov.uk

COURSEWORK TASKS

1 Add to this hazard prompt list:

Slips and trips; lifting; electricity; radiation; stress; transport; chemicals; computer screens; biohazards ....

(Your list will help you when writing your safety report for your coursework investigation.)

2 You have probably been sunburnt or seen people who have.

a) How does '***time of exposure***' increase your ***risk*** of sunburn?
b) What ***control measures*** can be used to avoid being burnt?
c) What ***emergency action*** can you take if you get sunburnt?

2.4 Real life – it's not just the fire

LEARNING OBJECTIVES

1. Why is fire safety at work significant?
2. When do we use fire doors, fire extinguishers and automatic sprinkler systems?

Fighting a major fire at a warehouse

Think about how this company should avoid such a disaster in the future.

A smoke alarm

Sprinkler facts

- A sprinkler costs less than a carpet.
- Sprinklers are completely automatic. They work by themselves and can stop heat and smoke from trapping people.
- Only the sprinklers over a fire open. All the others stay shut.
- In a fire you are 20 times more likely to die in a building without a sprinkler.
- Smoke is the main cause of death and damage to property in fires.
- In buildings protected by sprinklers, 99% of fires are controlled.
- Fire fighters often use 10 000 times more water from hoses to do the same job as a sprinkler.

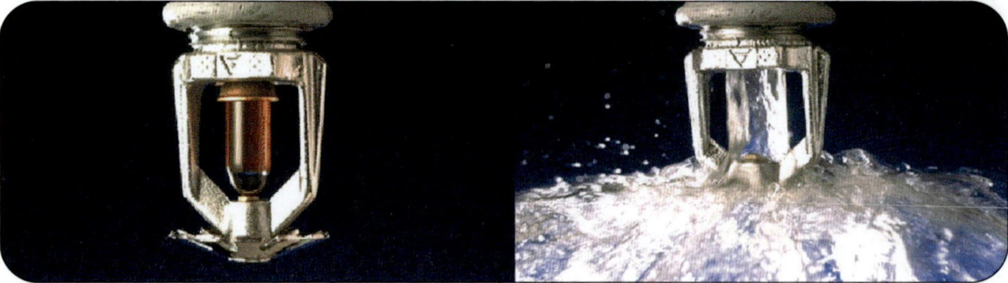

The liquid-filled glass bulb shatters at 65 °C. The sprinkler head opens and sprays water onto the fire.

A fire blanket

a) How do sprinklers reduce the amount of water needed to put out a fire?
b) How can sprinklers also reduce the amount of smoke?
c) Why does less damage occur in buildings that have sprinkler systems?

Sprinklers are not the answer for every place of work. In schools, for example, we have fire alarms, fire extinguishers and fire blankets.

d) Why are these preferred, rather than sprinklers? (Think about old buildings and vandalism.)

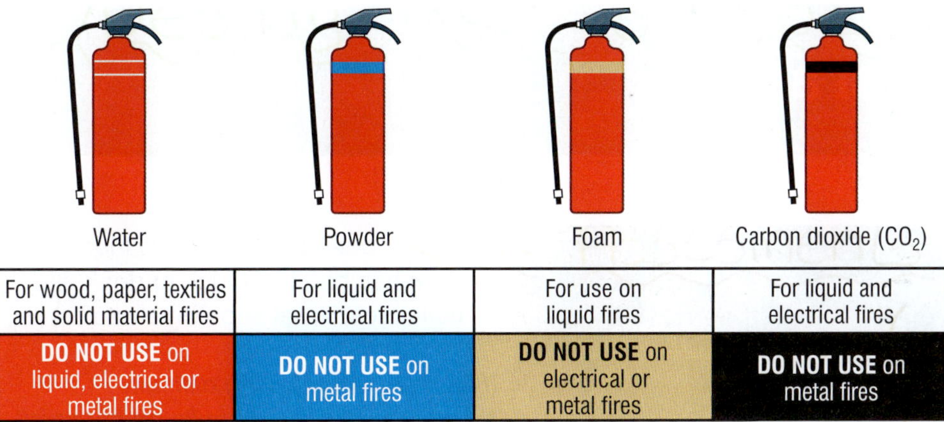

Water	Powder	Foam	Carbon dioxide (CO_2)
For wood, paper, textiles and solid material fires	For liquid and electrical fires	For use on liquid fires	For liquid and electrical fires
DO NOT USE on liquid, electrical or metal fires	**DO NOT USE** on metal fires	**DO NOT USE** on electrical or metal fires	**DO NOT USE** on metal fires

Types of fire extinguisher

Classes of fire

- Class A: ***solids*** such as paper, wood, plastic, etc.
- Class B: ***flammable liquids*** such as paraffin, petrol, oil, etc.
- Class C: ***flammable gases*** such as propane, butane, methane, etc.
- Class D: ***metals*** such as aluminium, magnesium, titanium, etc.
- Class E: fires involving ***electrical apparatus***.

Types of extinguisher

Water: Works by cooling material to below its ignition point. Used for Class A fires. Not suitable for Class B (liquids) or Class E (electricity) fires.

Dry powder: Works by smothering (by excluding oxygen) and knocking down flames. For Class A, B and E fires, but best for fires involving flammable liquids. *It can be dangerous to extinguish a gas fire without first turning off the gas supply.*

Foam: Works by forming a foam blanket to smother the fire and stop combustion. For Class A and B fires. (Not recommended for home use or electrical fires, but safer than water.)

Carbon dioxide (CO_2): Carbon dioxide smothers the fire by displacing oxygen. Is ideal for fires in electrical apparatus, and will extinguish Class B (flammable liquid) fires. *Carbon dioxide can be dangerous in confined areas.*

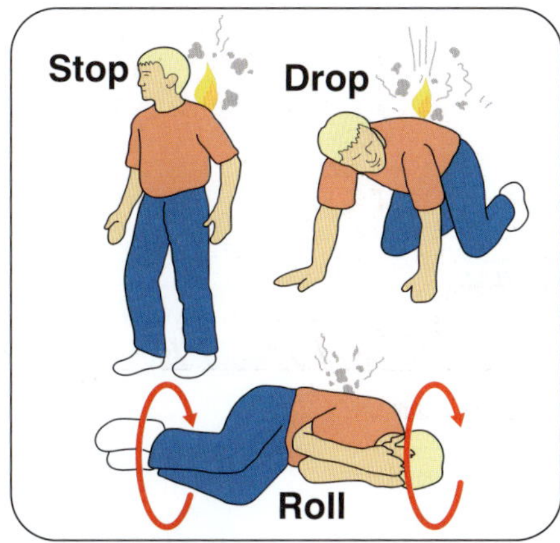

What should you do if your clothes catch fire?

Fire doors form a barrier to stop fire spreading. They must be kept shut but not locked!

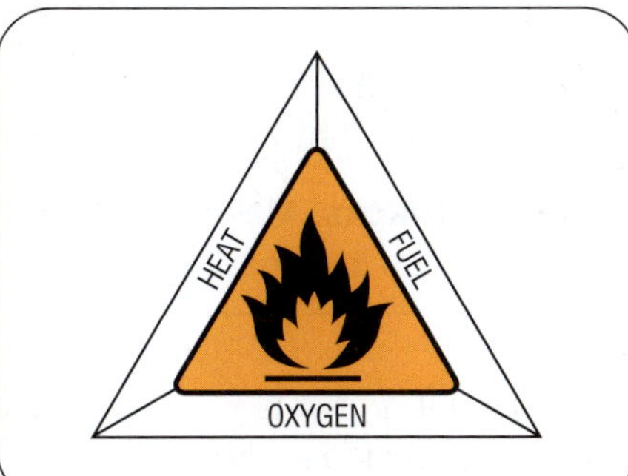

The fire triangle

DID YOU KNOW?

Cigarettes burn at 700 °C.

Every 3 days, someone dies from a fire caused by a cigarette.

COURSEWORK TASKS

In your coursework report include sections about fire safety in your:

- school or college.
- chosen scientific workplace.

2.5 First aid

LEARNING OBJECTIVES

1 What do we mean by basic first aid in labs?
2 When is it dangerous to give first aid?
3 How can we obtain first aid qualifications?

Sam after the party!

a) You've been to a party with friends. Sam has had too much to drink and collapses. What should you do?

If Sam is clearly having difficulty breathing, she could choke on her vomit. Call 999 for advice.

If Sam comes round, give her water or fruit juice to drink. (Coffee won't sober her up. It just increases the rate at which alcohol goes round her body.)

Sam was glad her friend knew what to do.

An accident happens. Someone is injured. It's important to treat them as soon as possible.

Imagine an accident in a science laboratory. It is important to know what to do while waiting for a qualified first-aider. The following advice covers common laboratory accidents:

Chemical splashes in the eye. Wash the eye under tap water for at least 10 minutes. The flow should be slow and the eyelid should be held back. Afterwards, take the casualty to hospital.

Chemical splashes on the skin. Wash the skin until all traces of the chemical have disappeared. Remove clothing as necessary.

Chemicals in the mouth, perhaps swallowed. Do not wash out the mouth, as the chemical may be swallowed. After the ***first-aider*** washes the mouth out, take the casualty to hospital.

Burns. Cool under gently running water until first aid arrives.

Toxic gas. Sit the casualty down in the fresh air.

Clothing on fire. Smother by pushing the casualty to the ground. Spread the laboratory fire blanket on top of the flames.

Asthma attacks. Make sure that their personal medication is available, so they can use it immediately.

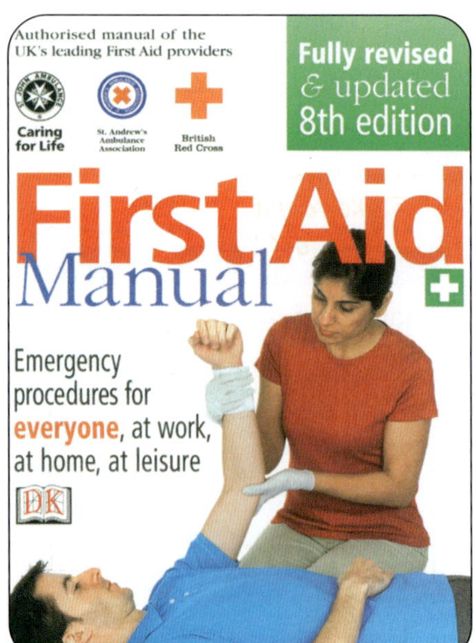

Most people who train in first aid use this manual (see www.dk.com/firstaidmanual). You can train with St John Ambulance (www.sja.org.uk) and British Red Cross (www.redcross.org.uk).

b) Research questions. What should you do if a friend gets:
   i) an electric shock?
   ii) a bad cut?
c) Can you use rubber gloves to move an electric shock casualty?
d) Why don't you take large objects out of a bad cut?
e) Students sometimes wipe their eyes and transfer chemicals on their hands into their eyes. What ways do you recommend to reduce these accidents?

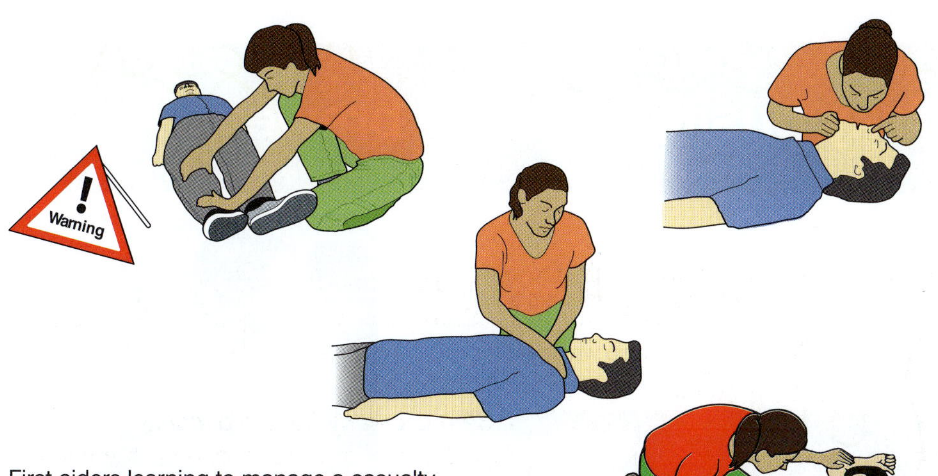

First-aiders learning to manage a casualty

The recovery position

First-aiders begin by assessing the ***danger***:

- Are you or the casualty in any danger?

They continue with ***response***:

- Does the casualty respond to you?

Then you have the ***ABC***:

- ***Airway, Breathing, Circulation.***

DID YOU KNOW?

- First aid myth – Nearly 1 in 5 parents surveyed said: "The best thing to put on a burn is butter or ice."
- Butter's useless. It's painful to remove it in hospital.
- Ice should not be used. It can produce a cold burn on top of the hot burn.

Fact – Place the burned area under cold water for 10 minutes, to reduce pain and distress. Then wrap it in cling film to reduce the risk of infection. Allow the area to cool more. If the burn is bigger than a 50p piece, go to a hospital.

COURSEWORK TASKS

1 Include sections in your coursework about the person responsible for first aid in your school and in your chosen scientific workplace. Which organisation did they receive their qualification from?

2 The table shows some common laboratory injuries. Refer to a First Aid Manual and design a first aid poster for your school or college. The poster must give information for each injury. Check your poster design with a qualified first-aider.

Injury	Your aims	First aid steps	Risks to your safety Cautions / Warnings
Minor burns / scalds			
Chemical burn			
Inhalation of fumes			
Swallowing chemicals			
Electric shock			
Cuts and grazes			
Foreign object in the eye			
Chemical in the eye			

3 Find out why it is useful to have a first aid qualification.

2.6 Risky business at the brewery

LEARNING OBJECTIVES

1 Where are there potential hazards in scientific workplaces?
2 When are health and safety regulations important in scientific workplaces?
3 How can risks arise?

The cleaner in the photo needs a permit to enter a fermentation vessel. For cleaning he wears a full body suit, with a harness attached. He has a carbon dioxide (CO_2) monitor on his waist. He also wears a motion sensor. If he stops moving, an alarm goes off.

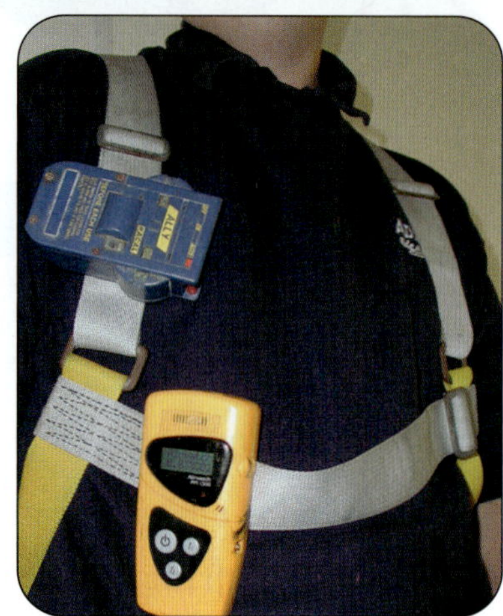

Why might it be risky to clean out one of these containers?

Why monitor carbon dioxide?

BACKGROUND SCIENCE

To make beer we use **yeast** to break down sugar to make alcohol. Scientists call this **fermentation**. They describe the process as **anaerobic respiration** because it doesn't need oxygen.

Glucose (sugar)	→	ethanol (alcohol)	+	carbon dioxide
$C_6H_{12}O_6$	→	$2C_2H_5OH$	+	$2CO_2$

The froth on top of the beer is filled with carbon dioxide gas. **Carbon dioxide** is more dense than air. It falls to the bottom of the fermentation vessel when you pump the beer out of the container.

The cleaner sprays sodium hydroxide (NaOH) on the walls of the vessel. This kills any bacteria that might start to grow there. Bacteria will ruin the flavour of the beer. Sodium hydoxide is an alkali. It is caustic and kills bacteria by attacking their protein.

a) How does the full body suit protect the cleaner?
b) Why might the cleaner have too little oxygen when breathing?
c) What emergency action could we take in the event of an accident?

The brewery are replacing their old fermentation vessels. The new ones are sealed containers. These vessels contain automatic sodium hydroxide spraying heads.

d) How are the newer vessels safer?

An automatic sodium hydroxide spraying head in a fermentation vessel

What happens when carbon dioxide levels increase?

0.04% The approximate amount of CO_2 naturally in the atmosphere

0.5% The normal International Safety Limit

0.7% Workers have 20 minutes to evacuate the area

1.2% Workers leave the area immediately

1.5% The normal Short Time Exposure Limit

3%

You will be breathing at twice your normal rate. Your heart rate and blood pressure increase. You feel dizzy, your head aches and your hearing is impaired.

5%

You breathe much faster and feel a choking sensation. You quickly tire and become confused.

People move heavy objects in the brewery. This is also a hazard. Using fork-lift trucks reduces the risk to workers. However, ***manual handling*** is still needed. The management provides training in safe lifting. Its rules are:

- Face the way you need to move.
- Bend your knees and lift with a straight back.
- Hold the load close to your body.

These rules are strictly enforced.

A worker not following the safe lifting rules gets a warning. *"You've heard it before: Lift with your legs, not with your back. So how come you bent your waist lifting that case of beer?"*

Back injury

e) What would happen next if a worker is seen not lifting safely. (Is it unreasonable for the worker to get the sack?)

f) Discuss why there may be a CO_2 risk in:
   i) mushroom farms and greenhouses
   ii) soft drink production
   iii) residential homes for the elderly.

COURSEWORK TASKS

1 Describe potential hazards in scientific workplaces due to:
   a) careless behaviour
   b) not using equipment properly
   c) not using protective and safety equipment
   d) not following correct procedures.

2 In your portfolio write about:
   a) Health and safety checks.
   b) Risk assessments for hazardous materials and procedures.
   c) What to do to prevent accidents.

2.7 Health warning

a) What is the difference between a hazard and a risk?

Hazards and risks of a new game

Imagine that a new game takes off as the next school craze. Your head teacher has had a complaint from a school governor. So your Applied Science class has been given the task of completing a risk assessment for the game.

Your head teacher warns you that different people must not blow into the same balloon. "I don't want anyone passing on their germs," are the head teacher's stern words. "We've just had an outbreak of 'flu and you've all heard of MRSA, haven't you?"

 Discuss your ideas of what should be included in the risk assessment.

Complete a risk assessment form for the game:

Hazards	Risks	Control measures	Emergency action

Fighting the superbug

You have probably seen newspaper headlines such as 'Superbug killed my husband' or 'Dirty hospital disgrace.'

But what is MRSA?

MRSA is a bacterium, which causes infections. It is a 'superbug', because MRSA is resistant to antibiotics. MRSA is short for Multiple Antibiotic-Resistant *Staphylococcus aureus.*

Staphylococcus aureus is a common bacterium. It lives on the skin of many healthy people without them even knowing it. People in hospital tend to be old, sick, weak and at risk of infections. While their defences are low, MRSA is a potential killer.

Sadly, infections of MRSA in hospitals lead to approximately 5000 deaths every year.

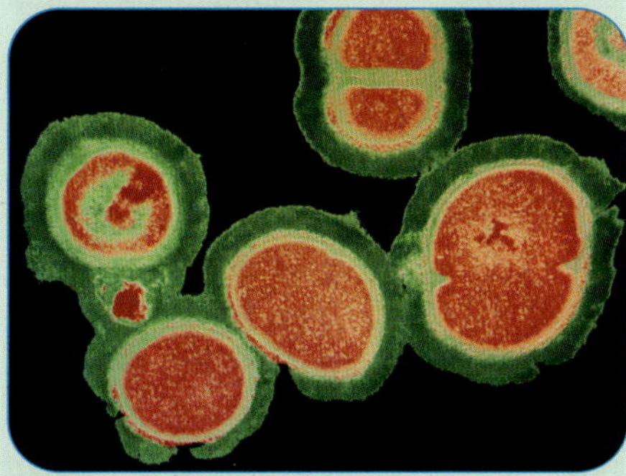

Staphylococcus aureus

b) What is an antibiotic?
c) How could an infection of MRSA on the skin get into the bloodstream? (**Clue:** think about a wound.)
d) Why is an infection in the blood more dangerous than one on the skin?

What do doctors say about MRSA?

Hygiene has always protected patients from dangerous diseases.

- Moving from patient to patient, doctors and nurses must always wash their hands.
- Visitors must also use the 'alcohol-based hand-rubs'. These are designed to reduce the number of micro-organisms on their hands.
- To be safe, hospitals must also be clean.

'Cleanyourhands' campaign logo

Hospitals are currently spending £1 billion each year fighting infections such as MRSA. However, in school, healthy young people are not in danger of MRSA infections.

"Whether a dirty ward rather than a dirty hand is a breeding ground for *Staphylococcus* is a matter of debate". What do you think?

e) Why does MRSA affect me?

Food science

This scientist is ensuring that foods are of good quality by testing them for microbes, and by providing nutritional information

What you already know

Here is a quick reminder of previous work that you will find useful in this chapter:

- A healthy diet contains the right balance of the different foods you need and the right amount of energy. A person is malnourished if their diet is not balanced. This may lead to a person being too fat or too thin. It may also lead to deficiency diseases.
- In the developed world too much food and too little exercise are leading to high levels of obesity and diseases linked to excess weight. These include arthritis (worn joints), diabetes (high blood sugar), high blood pressure and heart disease.
- Some people in the developing world suffer from health problems linked to lack of food. These include reduced resistance to infection and disease.
- Cholesterol is a substance made by the liver and found in the blood. The amount of cholesterol produced by the liver depends on a combination of diet and inherited factors. High levels of cholesterol in the blood increase the risk of disease of the heart and blood vessels.
- Too much salt in the diet can lead to increased blood pressure for about 30% of the population.
- Processed food often contains a high proportion of fat, salt and additives.
- Processed foods may contain additives to improve appearance, taste and shelf life. These additives must be listed in the ingredients and some permitted additives are given E-numbers.
- Chemical analysis can identify additives in foods. Artificial colours can be detected and identified by chromatography.

From Key Stage 3:

- To remain healthy you need a balanced diet containing carbohydrates, proteins, fats, minerals, vitamins, fibre and water, and know about foods that are sources of these.
- Plants need carbon dioxide, water and light for photosynthesis, and produce biomass and oxygen.
- How to summarise photosynthesis in a word equation.
- Nitrogen and other elements, in addition to carbon, oxygen and hydrogen, are required for plant growth.

RECAP QUESTIONS

1. What is a balanced diet?
2. Name three examples of foods that are rich in:
   a) proteins,
   b) carbohydrates,
   c) vitamins and minerals.
3. Eating too much fat is harmful to your health, but why are fats an important part of your diet?
4. What do diabetics constantly need to monitor?
5. Why is high blood pressure damaging to your health?
6. What is the word equation for photosynthesis?
7. Name the four main mineral elements plants need for healthy growth.
8. Farmers exploit the process of photosynthesis in their greenhouses to make plants grow as quickly as possible. Describe the optimum conditions for plant growth.

Making connections

Food scientists

- Check that a food contains the same ingredients as listed on the label.
- Check for microorganisms that could cause food poisoning.
- Detect the presence of additives, which may have been added to a food to enhance its flavour, texture or appearance.

Useful microorganisms

Some microorganisms benefit humans by producing useful food and drink products. Bacteria are used in making cheese and yoghurt. We also use the fungus yeast to make beer, bread and wine.

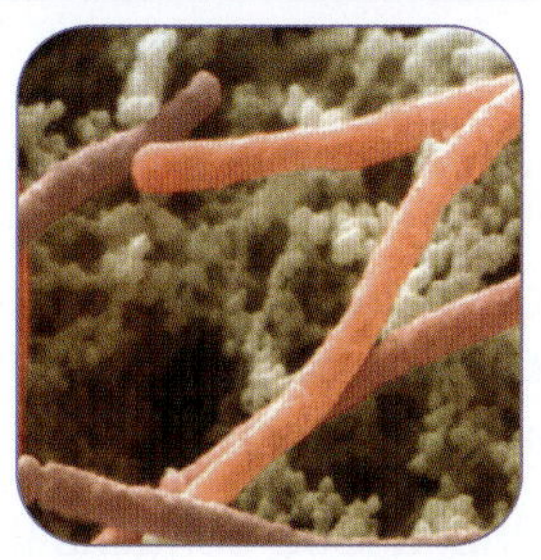

Dieticians

A dietician studies a person's diet by analysing their nutritional intake. They can then give advice on how to eat a healthier diet. Dieticians also use their expertise in hospitals to prepare special diets for people who suffer with allergies. They also help people with food-related medical conditions like diabetes.

Intensive farming

Intensive farming involves the use of chemicals to protect plants and animals from disease and pests, and to enhance growth. These chemicals maximise yields from an area of land. However, the chemicals can enter food chains, killing organisms they were not intended for. The use of large machinery can also damage the environment.

Healthy diet

A healthy diet contains lots of fruit and vegetables. Most of your energy should come from starchy foods such as pasta and rice. You should have a low intake of fat (especially saturated fat), salt and sugar.

Organic farming

Organic farmers only use a small number of artificial chemicals allowed by law. However, food yields are generally much smaller than those produced by more intensive means.

Food poisoning

Some microorganisms that live in food products can be harmful to human health, causing food poisoning. This can lead to some unpleasant symptoms such as sickness and diarrhoea. Storing and cooking food properly usually avoids this.

ACTIVITY

Lots of people are involved in growing your food and checking it is safe to eat. They advise us on the types and amounts of each food we should eat so that we grow and stay healthy. Choose one aspect of food production to research. Using this information produce a short PowerPoint presentation explaining:

- How we grow the raw materials for the food (organically or intensively).
- How we turn the raw materials into the final product.
- The nutritional content of the product produced and the beneficial or harmful effect it has on your body.

3.1 Introduction to food science

LEARNING OBJECTIVES

1. What is the role of a dietician?
2. What is the role of a food scientist?
3. How is food produced commercially?

The Food Standards Agency was set up in 2000 to protect the public's health and consumer interests in relation to food. Food scientists and dieticians work within the Agency to promote good eating habits. They ensure that your food is safe and that it is labelled correctly.

The role of a ***dietician*** is to study a person's diet. They record what they eat and analyse their intake. They can then recommend how individuals can eat a healthier diet. Dieticians also use their expertise in hospitals. They prepare special diets for people who suffer from allergies or have food-related conditions like diabetes.

a) How can a dietician help an obese person to improve their health?

A healthy diet contains lots of fruit and vegetables. You should get most of your energy from starchy foods such as pasta and rice. You should also have a low intake of fat (especially saturated fat), salt and sugar.

Food scientists check that a food contains the same ingredients as the information manufacturers put on the label. They test the content of foods, from the raw materials used through to the final product.

Food scientists use chemical tests to detect the presence of fats, sugars and protein. They check for the presence of microorganisms that could potentially cause food poisoning. They can also detect additives that may have been added to a food to improve its flavour, texture or appearance.

This dietician is advising a man about which foods to avoid to lose weight and which foods are rich in nutrients that he needs to maintain a healthy lifestyle

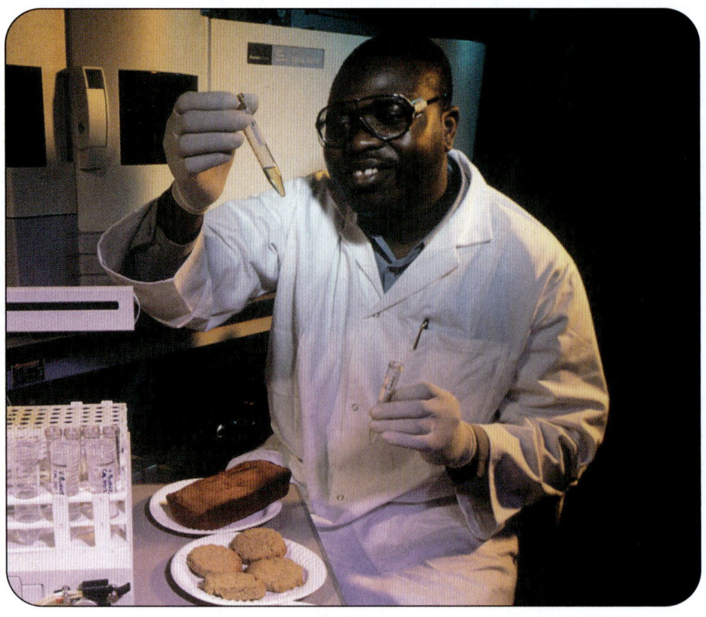

This food scientist works in quality control analysis, checking the nutritional value of a food product

b) Name three substances that a food scientist can test in a food product.

Harmful microorganisms in our food can cause food poisoning. Strict hygiene and cleaning procedures must be followed to prevent food contamination. Once you have prepared food it must then be stored properly to prevent it 'going off'.

Some microorganisms can be used to benefit humans by producing useful food and drink products. We use bacteria in the production of cheese and yoghurt. The fungus yeast is used to make beer, bread and wine.

c) Name some food products that we can make using microorganisms.

The majority of the world's food is produced by ***intensive farming***. This involves the use of a range of chemicals:

- to protect plants and animals from disease,
- to reduce the problem of predators and competition from other plants and animals,
- to enhance growth.

These chemicals maximise yields from an area of land. However, the disadvantages are:

- they can enter food chains killing animals they were not intended for. They can also kill beneficial insects.
- the use of large machinery can damage the environment.

d) Name an advantage and disadvantage of intensive farming.

Organic farmers only use a small number of artificial chemicals allowed by law. This results in a more natural product, and generally provides better living conditions for animals. However, yields are generally much smaller. This results in the products being more expensive than those produced by more intensive means.

e) Name an advantage and disadvantage of organic farming.

This is a Public Health Laboratory. One of its roles is to detect outbreaks of food poisoning, such as salmonella associated with eggs and poultry. This microbiologist is removing a sample from a defrosted chicken to check for the presence of bacteria.

This farmer is spraying his maize crop with pesticides. These kill pests that may damage the crop and reduce its yield. Many pesticides are harmful to humans and the environment. Protective clothing should be worn during spraying, and traces of the chemicals are monitored in food products.

SUMMARY QUESTIONS

1 Look at this food label:

> **ORGANIC**
> **CHEDDAR CHEESE**
> Free from artificial colourings and flavourings.
> Keep refrigerated.
> Packaged in a protective atmosphere:
> once opened best consumed within 3 days.

a) What type of farming method was used to produce this product?
b) Does this food contain any additives?
c) Where must you keep this product to prevent unwanted bacteria growing in it?
d) What type of microorganism was used to produce this food?

2 What are the main differences between organic and intensive farming?

3 What is the difference between a dietician and a food scientist?

KEY POINTS

1. Dieticians study a person's diet and make recommendations as to how they can make changes to improve their health.
2. Food scientists monitor the contents of food products checking their nutritional content and the presence of unwanted microorgansims.
3. There are two methods of food production: organic and intensive farming.

3.2 Nutrients

LEARNING OBJECTIVES

1 What are nutrients?
2 What are nutrients used for in your body?
3 Why is fibre important in your diet?

DID YOU KNOW?

If you were to eat too many carrots you would turn orange, and eating too much asparagus turns your urine bright yellow! Make sure you eat a balanced diet.

Doctors and nutrition experts constantly say that, in order to remain healthy, you have to eat a nutritious balanced diet. What does this mean, and why is it important?

Nutrients are essential elements or compounds that your body needs to carry out the vital life functions of respiration, movement, growth and repair of body tissue. They are classified into six nutrient groups:

- carbohydrates
- fats
- proteins

} These nutrients provide energy for the body.

- vitamins
- minerals
- water

} These nutrients are essential for the body to function normally.

a) What is a 'nutrient'?
b) What are the six types of nutrient needed by the body?

Carbohydrates can be found in a wide range of foods

Carbohydrates

Carbohydrates are your main source of energy. There are two main types of carbohydrate in foods:

- ***Simple carbohydrates*** (also called simple sugars) are found in refined sugars, like caster sugar. They are also found in more nutritious foods like fruit and milk. They are the major source of energy for the body.
- ***Complex carbohydrates*** (also called starches) have to be broken down by the body into simple sugars. Starches include grain products, such as bread and pasta.

c) Which type of carbohydrate should you eat if you need a sudden energy boost?

Fats

Fats provide you with a store of energy and help to keep you warm, by providing a layer of insulation under your skin. Fat covers your vital organs, like your kidneys and heart. This protects them from damage. They also provide a source of fat-soluble vitamins (see page 33), which are needed for healthy growth and development.

Fats can be obtained from a number of different sources

There are two main types of fat:

- **Saturated fats**, which come mainly from animals.
- **Unsaturated fats**, which come from plants.

If you eat too much saturated fat, the lining of your arteries may become thickened. This makes them narrower, so your heart has to work harder to pump the blood through them. Also there is a greater risk of you having a heart attack.

Unsaturated fats are considered healthier than saturated fats. That's because they do not occur with a substance called **cholesterol**, which is linked to heart disease.

d) What type of fat is found in cheese?

Proteins

Proteins are needed for repairing body tissues, growth and energy. Your muscles, organs, and immune system are made up mostly of proteins.

Your body uses the protein you eat to make specialised protein molecules. These molecules carry out specific jobs in your body. For example, proteins are used to make haemoglobin, a major component of red blood cells.

Fibre

Fibre is a form of carbohydrate, but it is ***not*** classed as a nutrient because it doesn't provide us with any goodness. We don't digest fibre, but it is an essential component of your diet because it adds bulk to your food. This means that waste can be pushed out of the digestive system more easily.

Fibre also absorbs poisonous waste and makes you feel full, which helps to stop you overeating. A high-fibre diet is very effective in preventing constipation.

e) Name three advantages of eating a fibre-rich diet.

SUMMARY QUESTIONS

1 Copy and complete the following summary table:

Food group	Examples of food	Role in the body
Carbohydrates		
Fats		
Proteins		
Fibre		

2 Why would a doctor suggest to a patient who is constipated that they should eat more cereal?

3 A packet of sweets and a glass of milk both contain simple sugars. Why is it better to get your simple sugars from milk rather than sweets?

4 Vegetarians have to be very careful that they eat enough protein. Design a vegetarian meal that contains carbohydrate, protein, fat and fibre.

GET IT RIGHT!

Questions relating to balanced diets and food groups are very common. Know the main food groups, their role in the body, and examples of foods that contain them.

Fish and meat are examples of foods rich in protein

Fibre is found in cereals, fruit and vegetables, nuts, seeds and beans

KEY POINTS

1 Nutrients are essential elements or compounds.

2 Nutrients are needed to carry out the vital life functions: respiration, movement, growth and repair of body tissue.

3 Fibre is important for aiding digestion as it prevents you from overeating and absorbing poisonous waste.

3.3 Vitamins and minerals

LEARNING OBJECTIVES

1. What is the difference between a vitamin and a mineral?
2. Where do we get minerals and vitamins from?
3. What are vitamins and minerals used for in the body?

Health food companies bombard us with adverts telling us to buy multi-vitamins and minerals. Is it necessary for you to supplement your diet, or can you get them naturally?

Vitamins

Vitamins are chemical compounds that your body needs in small amounts so that it can grow, develop and function normally. Your body needs 13 different vitamins. You can only make four, so vitamins are an essential part of a healthy diet.

a) What is a vitamin?

Vitamins perform many functions in your body. For example:

- ***Vitamin A*** helps to maintain healthy eyesight and skin. It also keeps mucous membranes (like the nose and mouth) free from infection. Spinach and carrots are rich in vitamin A.
- ***Vitamin B*** vitamins (a group of vitamins) are involved in releasing energy from carbohydrate foods and nerve functions. Nuts, dark green vegetables, yeast extract and liver are a good source of vitamin B.
- ***Vitamin C*** is used to maintain your immune system, help your body absorb iron and to maintain your skin and the linings of your digestive system.
- ***Vitamin D*** is needed to form healthy teeth and bones. It also helps to absorb calcium and phosphorus. Vitamin D can be found in oily fish and fortified milk.
- ***Vitamin K*** aids blood clotting. It is found in dark green vegetables.

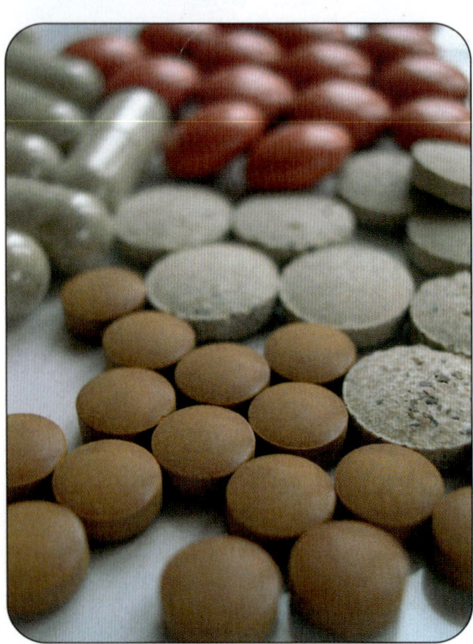

There is a wide range of vitamin and mineral supplements available

Fruits and vegetables are an ideal source of vitamin C

DID YOU KNOW?

Exposure of the skin to sunlight enables the body to produce sufficient amounts of vitamin D.

b) Name some examples of foods that contain vitamin C.

Vitamins A, D, E and K are ***fat-soluble***. This means they can be stored in your body in fatty tissues and the liver until you need them. B vitamins and vitamin C are ***water-soluble***. These vitamins must be used by your body straight away, otherwise they will get passed out of your body in urine.

c) What is the difference between a fat-soluble and a water-soluble vitamin?

Some food, like bread, cereals and milk, are ***enriched*** with vitamins and minerals. This means that they have had vitamins and minerals added to them.

Minerals

Minerals are substances that must be present in your diet for you to remain healthy. They are needed for several body functions, including transmitting nerve signals, maintaining a normal heartbeat, and producing hormones. There are at least 20 minerals needed by your body.

Four important examples are:

- ***Iron***, which helps your body to make haemoglobin. Haemoglobin makes blood red and is responsible for transporting oxygen around the body. Iron-rich foods include spinach, broccoli and liver.
- ***Phosphorus***, which is involved in releasing energy from your food. It can be found in meats, fish and eggs.
- ***Calcium***, which is important for healthy teeth and bones. Dairy products are good sources of calcium.
- ***Zinc***, which is a part of many essential enzymes, and plays an important role in helping to heal wounds. It is found in meat, liver and seafood.

d) Which vitamin and mineral do we need for healthy bone growth and development?
e) What are 'enriched' foods?

Dieticians and other medical professionals say that the best way for a healthy person to obtain the vitamins and minerals they need is to eat well-balanced meals. If a diet is missing any important components, it is said to be 'mineral-deficient' or 'vitamin-deficient'. This can lead to serious health implications.

SUMMARY QUESTIONS

1 Copy and complete the following sentences using these words:

healthy **minerals** **tiny** **vegetables**

The main sources of vitamins and . . . are fruits and . . ..
These compounds are only needed in . . . amounts, but are essential for remaining . . ..

2 What is the difference between a vitamin and a mineral?

3 If a woman loses lots of blood during childbirth, her doctor may prescribe a mineral supplement. What mineral would this supplement contain? What foods should she eat to gain this mineral from a natural source?

4 Why does vitamin C need to be replaced on a daily basis, whereas vitamin K can remain in the body for a longer period of time?

KEY POINTS

1 Vitamins are chemical compounds, whereas minerals are chemical elements.
2 Vitamins and minerals are present naturally in your diet – good sources are fruits and vegetables.
3 Vitamins and minerals are needed by your body to grow, develop and function normally.

3.4 Vitamin deficiencies

LEARNING OBJECTIVES

1 What do we mean by a 'vitamin deficiency'?
2 What effect does a lack of vitamins have on your body?

Nutritional scientists have provided guidance on the amount of each vitamin you need each day. This is known as the ***Recommended Daily Allowance (RDA)***. If you do not take in the RDA of a vitamin through food or drink you are said to have a **vitamin deficiency**. If this continues over a period of time, your health will be affected. Vitamin deficiencies can cause unpleasant symptoms and diseases.

a) What do we mean by the term 'RDA'?
b) Look at the label below. How much vitamin C and vitamin D should you take in every day?

Nutrional information

Typical Values	Per Daily Dose	%RDA*
Vitamin A	800µg	100
Vitamin D	5.0mg	100
Vitamin E	10mg	100
Vitamin C	60mg	100
Thiamin (Vitamin B1)	1.4mg	100
Riboflavin (Vitamin B2)	1.6mg	100
Niacin (Vitamin B3)	18mg	100
Vitamin B6	2.0mg	100
Folic Acid	200µg	100
Vitamin B12	1.0µg	100
Biotin	0.15mg	100
Pantothenic Acid (Vitamin B5)	6.0mg	100
Vitamin K	30µg	-
Calcium	220mg	27.5
Phosphorus	40mg	5
Iron	14mg	100
Magnesium	60mg	20
Zinc	15mg	100
Iodine	150µg	100
Copper	1.0mg	-
Chloride	36mg	-
Chromium	25µg	-
Manganese	2.5mg	-
Molybdenum	25µg	-
Potassium	40mg	-
Selenium	25µg	-

*RDA means Recommended Daily Amount

Food supplements are intended to supplement the diet and should not be substituted for a varied diet. Free from Artificial Colours, Flavours and Preservative. For recycling purposes, this container is high density polyethylene and the cap is polypropylene.

Tablet Size

complete a-z
with multivitamins and minerals

help support your nutritional balance

90
one a day tablets

Best Before End Date: 05 2006 48

Labels like this are displayed on all multivitamin packets

DID YOU KNOW?

The World Health Organisation has estimated that 250 000 to 500 000 vitamin A-deficient children in developing countries become blind every year. Half of them die within 12 months of losing their sight. Health experts say that giving them vitamin A supplements twice a year can prevent the majority of these children from going blind. This treatment costs less than £1.

Vitamin A deficiency

Too little vitamin A can result in dry skin and mucous membranes – these line your throat, nose, mouth and lungs. This deficiency can also make it difficult for your eyes to adjust to dim light. This is called 'night blindness'.

Try having a raw carrot as a snack to increase your vitamin A intake.

c) What are the symptoms of vitamin A deficiency?

This student has fainted – he is suffering from anaemia

Vitamin B deficiency

Not eating foods that contain enough vitamin B can result in mouth sores and the degeneration of nerve cells. Vitamin B is also needed to make red blood cells. Without it you can develop a disease called 'anaemia'. If you are anaemic you do not have enough red blood cells. This means that your blood cannot carry enough oxygen, making you feel very tired and dizzy.

Try increasing your vitamin B intake by eating a handful of nuts.

Vitamin C deficiency

If you don't eat enough foods containing vitamin C, this may lead to soft and bleeding gums, slow-healing wounds, bruising and nosebleeds. This is because vitamin C is needed to form strong blood vessels. If the walls of your blood vessels are weak, they can rupture easily, allowing blood to escape. When this happens under the surface of your skin, it causes bruising.

Try having a glass of fruit juice every day to boost your vitamin C levels.

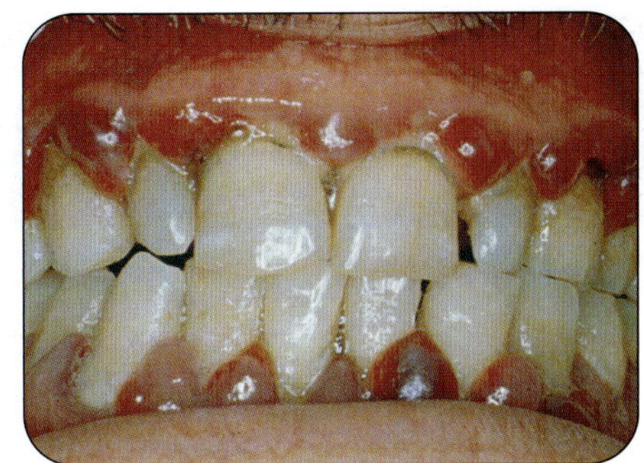

These are the teeth of a person suffering from scurvy – a disease caused by a vitamin C deficiency

GET IT RIGHT!

Do not write more than you need to. If a question asks you to 'describe', just write what happens. For example, a question may ask you to 'Describe the symptoms of vitamin C deficiency'. Your answer could say 'Soft and bleeding gums, slow-healing wounds, bruising and nosebleeds'. You do not need to say ***how*** a vitamin C deficiency causes these symptoms.

Vitamin D deficiency

A vitamin D deficiency in your body causes weak teeth and bones. Vitamin D is needed for your bones to absorb calcium. If this is missing, your bones will soften over time as calcium is lost from the skeleton. Soft bones are flexible, and can become deformed by the weight of your body. This disease is called rickets, and is most common when children are growing rapidly.

Try having a bowl of cereal with milk every morning. Many cereals are fortified with vitamin D, and milk contains plenty of calcium.

d) What is 'rickets'?

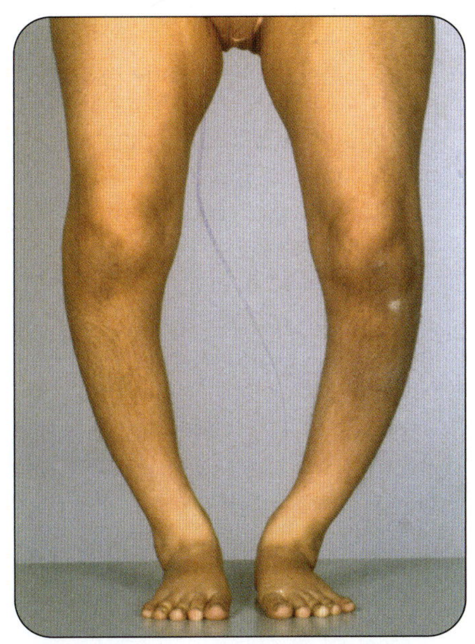

The bowed legs of a person with rickets

SUMMARY QUESTIONS

1 Match the vitamin deficiency to some of the symptoms it would produce:

Lack of Vitamin A	**weak teeth and bones**
Lack of Vitamin B	**anaemia and mouth sores**
Lack of Vitamin C	**dry skin and poor night vision**
Lack of Vitamin D	**bleeding gums and slow-healing cuts**

2 Is there any scientific truth in the saying 'Eating carrots helps you to see in the dark'?

3 Explain why a person suffering from anaemia often feels tired.

4 If you have a vitamin C deficiency, your gums are likely to bleed when you brush them. Explain why.

KEY POINTS

1 If you do not take in the Recommended Daily Allowance (RDA) of a vitamin through food or drink, you are said to have a 'vitamin deficiency'.

2 Vitamin deficiencies affect your health causing unpleasant symptoms and, in more severe cases, diseases.

3.5 How does your diet affect your health (1)?

LEARNING OBJECTIVES

1 Why do people's energy requirements change depending on their age, sex and job?
2 Why is eating too much saturated fat bad for your health?

'Fast food' has become a part of a busy lifestyle. That's because it is convenient, predictable, and fast. Nutritionists advise against eating fast food, as it is often high in sugar, salt, saturated fat and a number of different food additives. Eaten in large quantities, these food components can have a bad effect on your health both now and, more severely, later in life. This does not mean fast food harms you if eaten occasionally, but it must form only a small part of a balanced, healthy diet.

a) What are the main components of food that should be limited in your diet?

Energy in foods

Your body needs energy to function properly. The amount of energy you need depends on how old you are (this affects your growth rate), your body size and how active you are.

You need energy even if you are not 'doing' anything – you've still got to power your heart, lungs, maintain your body temperature and keep all the chemical reactions in your body going. ***Energy requirements*** are measured in kilojoules (kJ) per day.

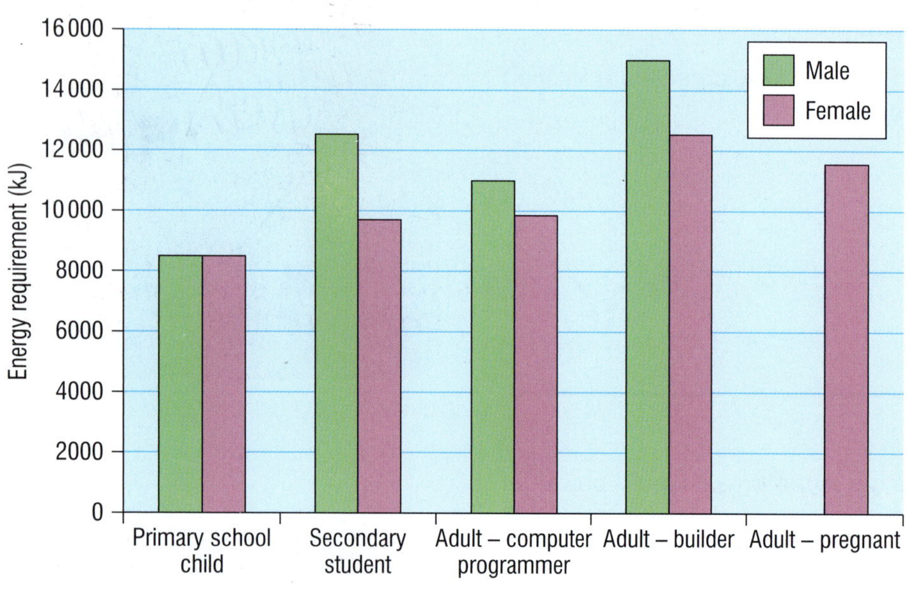

Daily energy requirements for different groups of males and females

b) How much energy do you need per day to function healthily?

If you do not take in enough energy in your diet, you will feel tired and lethargic. If you take in too much, you will put on weight. Energy-rich foods that are not used by the body are stored as fat under the skin. Fats contain the highest amount of energy, followed by carbohydrates (particularly sugars).

To ***maintain*** a desirable weight, men need about 11 000 kJ per day, and women around 9500 kJ per day. To lose weight, you must take in less energy than you burn. This means that you must eat foods with a lower energy content, increase your level of physical activity, or ideally both.

c) Why do pregnant women need more energy than other adult women?

DID YOU KNOW?

The human heart creates enough pressure, when it pumps blood around the body, to squirt blood a distance of about 10 m!

Saturated fat

Saturated fat is contained in animal products like butter and cheese (see page 31). Saturated animal fats can increase your blood cholesterol levels. Excess cholesterol in the body sticks to the walls of your arteries. Over time, this will narrow your arteries and can slow the flow of blood. The walls of the artery become rough, which can cause the blood to clot, blocking the vessel.

If your ***coronary arteries*** (the ones leading to your heart) get blocked, then it can cause heart disease. Partially blocked arteries can cause chest pains because not enough oxygen is getting to the heart muscle. This is called ***angina***. If the coronary arteries become blocked completely (a ***thrombosis***), it can cause a heart attack because the oxygen supply is cut off completely. If the heart doesn't start beating within a few minutes, the person will die.

d) How does eating saturated fat affect your blood vessels?

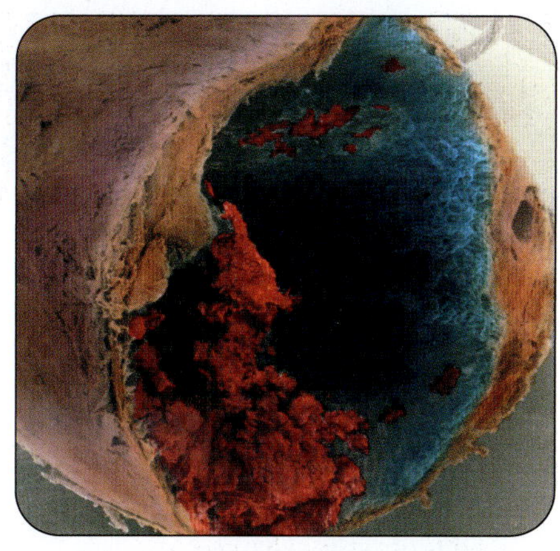

This photograph, taken with a microscope, shows a blood clot (red) in a coronary artery. To prevent this happening to you, dieticians recommended limiting your fat intake to 50–80 g per day.

PRACTICAL

Testing food products for the presence of saturated fat

Food scientists can determine how much saturated fat a food product contains by looking at its **iodine number**. This number is the mass of iodine (in grams) that can be absorbed by 100 g of a chemical substance. Olive oil, for example has an iodine number of 75 to 95, whereas butter has an iodine number of 25 to 40. The ***lower*** the iodine number, the ***higher*** the saturated fat content in a food.

Iodine in solution reacts with unsaturated fat in a food and loses its colour. To find out how much saturated fat there is in a food, iodine solution is **titrated** into the food being tested. The amount of iodine (in grams) that can just be decolourised by 100 g of the food gives the iodine number.

The titration equipment used to calculate the iodine number of a food product

SUMMARY QUESTIONS

1 Copy and complete the following sentences using these words:

amounts **balanced** **health** **saturated fats** **groups**

You should eat a . . . diet. This means that you should ensure that you take in each of the different food . . . , in the right . . .. Too much of any one thing (in particular. . . , sugar and salt) can cause a number of . . . problems.

2 Why is there a difference between the amount of energy needed per day by a manual worker and by an office worker?

3 Explain why an obese person is in greater danger of having a heart attack than someone who maintains a normal body weight.

KEY POINTS

1 The more active a person is, the more energy they require from their diet so that their body can function properly. You also need more energy when you are growing.

2 Eating too much saturated fat in your diet can narrow your blood vessels and cause heart disease.

3.6 How does your diet affect your health (2)?

LEARNING OBJECTIVES

1 Why is eating too much sugar on a long-term basis bad for your health?
2 Why should you restrict your salt intake?

The UK recommended intake for sugar is no more than 60 g of sugar per day. Check the food label to see if a product contains sugar – sucrose, glucose, maltose and fructose are all types of sugar.

The Food Standards Agency recommends that adults eat less than 6 g of salt a day. It is sometimes hard to tell if sodium is added to a food – check the label! Some common added forms include monosodium glutamate and sodium bicarbonate.

Sugar

Sugars are simple carbohydrates (see page 30). Sugars:

- Help improve the texture of food by keeping it moist.
- Give a golden caramel colour to some foods.
- Can extend a food product's shelf life.
- Are used to preserve fruits and make jams and marmalades.

Sugars provide instant energy for the body as they are absorbed quickly into the blood supply. However, they have little nutritional value. They taste nice, but are not very filling, so we may eat more than we need. This can lead to weight gain and eventually ***obesity***.

People who are obese are up to 80 times more likely to develop type 2 **diabetes** than people who maintain a healthy body weight. In this condition, a person's blood sugar level is too high as the body is not producing enough insulin, or the cells do not use insulin properly.

Insulin is a hormone that regulates blood sugar levels, by removing excess sugar from the blood. Diabetes can lead to serious complications such as heart disease, strokes and blindness.

a) What hormone is responsible for regulating blood sugar levels?

Salt

Sodium is an element that the body needs in order to function properly. It regulates blood pressure and blood volume and is critical for the functioning of muscles and nerves. The most common form of sodium in your diet is **sodium chloride (salt)**.

Sodium occurs naturally in most foods, and is even found in drinking water. Processed meats, such as bacon and sausages contain added salt. Fast foods, in particular, are very high in salt.

An intake of sodium above the RDA can cause ***high blood pressure***. This means that the pressure of blood in your arteries is too high. Over several years high blood pressure can damage your arteries, encouraging cholesterol to build up, making them narrower.

In general, the ***higher*** your blood pressure, the ***greater*** your risk of developing heart disease, kidney damage and having a stroke.

b) Why does your body need a small supply of salt?

As well as providing high levels of fat, sugars and salt, pre-prepared meals often contain colourings and flavourings. These too have been linked to health problems.

Food additives

There are over 3000 chemicals that food manufacturers can use to enhance the flavour, colour and texture of their products. Additives can provoke allergic or behavioural reactions. Sometimes these reactions occur within a few minutes, but others result from a build-up of an additive.

Government research has revealed that some permitted food colourings, which are used in many popular children's foods and fizzy drinks, can cause temper tantrums and disruptive behaviour in up to a quarter of toddlers. In older children the use of colourings has been linked with hyperactivity and exaggerating the symptoms of Attention Deficit Disorder. (For more health risks see pages 40–1.)

c) Why do manufacturers put food additives into their food?

Many takeaways contain food additives to enrich the colour and appearance of the food

Is this healthy?

Look at the photo of this ***obese*** man – being obese means you weigh at least 20% more than is healthy for you.

Discuss how you think he ended up like this. How is the food he is eating harmful to his health? What can he do to improve his health?

DID YOU KNOW?

If you are 10 kg overweight, you have nearly 5000 extra miles of blood vessels through which your heart must pump blood! What a lot of extra hard work for your heart!

SUMMARY QUESTIONS

1 Copy and complete the following table:

Food component	Recommended maximum daily intake	Health risks of eating too much of this component
Saturated fat		
Sugar		
Salt		

2 Make a list of 10 types of pre-prepared foods that contain high levels of sugar, salt, saturated fat or additives.

3 Why is high blood pressure a health risk?

4 Why do you think many schools have banned fizzy drinks from being sold in the canteen?

KEY POINTS

1 Excess sugar in your diet can lead to weight gain and eventually obesity. This increases your risk of developing diabetes.

2 Excess salt in your diet increases blood pressure. Over time this can result in heart disease, kidney damage and strokes.

3.7 What is in your food?

LEARNING OBJECTIVES

1 What are food additives?
2 Why are food additives used?
3 Can food additives damage your health?

Many foods have chemical additives added to them. Do you know what is in the food you are eating?

DID YOU KNOW?

If an E-number starts with a 1 it is a food colouring, with a 2 it is a preservative, with a 3 it is an antioxidant and with a 4 it helps the texture of the food.

Many colourings have been added to icing to produce these colourful biscuits

Many food manufacturers add chemicals to the food they are producing. These are called **food additives**. EU (European Union) legislation requires food additives to be clearly labelled in the list of ingredients, either by name or an **E-number**. An E-number means that it has passed safety tests and its use has been approved by the Food Standards Agency.

There are three main reasons why manufacturers use additives:

- To keep food fresh for longer periods of time – this increases its shelf life.
- To replace or enhance the flavour of food, which may be lost when the food is processed.
- To improve the appearance of foods, making it look more appetising.

a) Why do manufacturers add additives to food products?

Food additives can be grouped into six main groups: antioxidants, flavourings, colours, preservatives, sweeteners and thickeners.

Antioxidants

Antioxidants reduce the chance of fats and oils ***oxidising***. This oxidation causes food to change colour or go rancid. Antioxidants are used in a wide range of foods, including meat products, mayonnaise, bakery products and sauces. Vitamin C (ascorbic acid) is a widely used antioxidant.

Flavourings and flavour enhancers

Flavour enhancers are used to bring out the flavour in a wide range of foods, without adding a flavour of their own. Salt is not classified as a food additive, but is the most widely used flavour enhancer. Monosodium glutamate (MSG) is added to processed foods like pre-prepared meals, soups, and sausages. MSG can cause sensitive individuals to experience headaches, nausea and difficulty in breathing.

Flavourings are added to a wide range of foods, usually in very small amounts, to add a particular taste or smell to a food. Crisps are often manufactured using flavourings.

b) What is the difference between a flavouring and a flavour enhancer?

Colours

Food colourings are used to improve the appearance of foods. Some people think this is unnecessary and misleading. However, other people prefer their strawberry milkshake to appear pinker! Some food colourings are natural in origin. For example, curcumin is a yellow extract of turmeric roots. Others like tartrazine (E102) are artificial. This colouring is well known because it has been linked with hyperactivity in children.

Preservatives

Preservatives are used to help keep food safe for longer – sugar, salt and vinegar are still used to preserve some foods. Processed foods are likely to include preservatives, unless they have been frozen, canned, or dried.

Benzoic acid is commonly used. It is a naturally occurring preservative, found in many edible berries, fruits and vegetables. It can also be manufactured artificially. It is used to prevent bacterial and fungal growth in jam, fruit juices, yoghurts and soft drinks.

Concerns have been raised that the toxic nature of some preservatives could be damaging to your health.

c) Why are preservatives added to some foods?

GET IT RIGHT!

Make sure you know what the key terms mean. These are highlighted in bold in this book. Often the first part of an exam question asks you to explain a key term, and is worth 1 or 2 marks.

Sweeteners

To reduce their sugar intake some people choose to consume foods and drinks containing artificial **sweeteners**. These contain very little energy and are better for your teeth. Artificial sweeteners include aspartame and saccharin.

Thickeners

This group of additives contains **emulsifiers**, ***stabilisers*** and **thickeners**. Emulsifiers help mix together ingredients that would normally separate like oil and water. Stabilisers then prevent them from separating again. Both are used in making low-fat spreads. Thickeners (e.g. starch) and gelling agents (e.g. pectin) are used to thicken food to improve their texture.

Chocolate contains emulsifiers

d) What are the six main classes of food additives?

SUMMARY QUESTIONS

1 Complete the following sentences using these words:

unpleasant **chemicals** **shelf life** **additives**

Food . . . are . . . added to food when it is manufactured. They improve the colour, taste and . . . of food products. However, some have . . . side effects.

2 Why are flavour enhancers added to foods?

3 Produce a table to summarise the advantages and disadvantages of using different food additives.

4 By comparing the potential benefits and side effects of using artificial sweeteners, do you think that diet fizzy drinks are better for you than drinks containing sugar?

KEY POINTS

1 Food additives are chemicals that are added to food products.
2 Food additives keep food fresh for longer periods of time, flavour or enhance the taste of foods and can improve the appearance of foods.
3 Food additives have been linked to hyperactivity. Some cause unpleasant side effects and others have a toxic effect.

3.8

How can you tell what is in the food you are eating?

LEARNING OBJECTIVES

1 What information can we find on a food label?
2 What is the difference between 'sell by', 'best before' and 'use by' dates?

All food and drink products have a food label. These provide you with information about the ingredients the food contains, including any additives. They also give you advice on how you should store the product and the nutritional value it gives you. Food manufacturers have to follow rules that protect us from false claims or misleading descriptions. There are clear guidelines on what labels can and can't show.

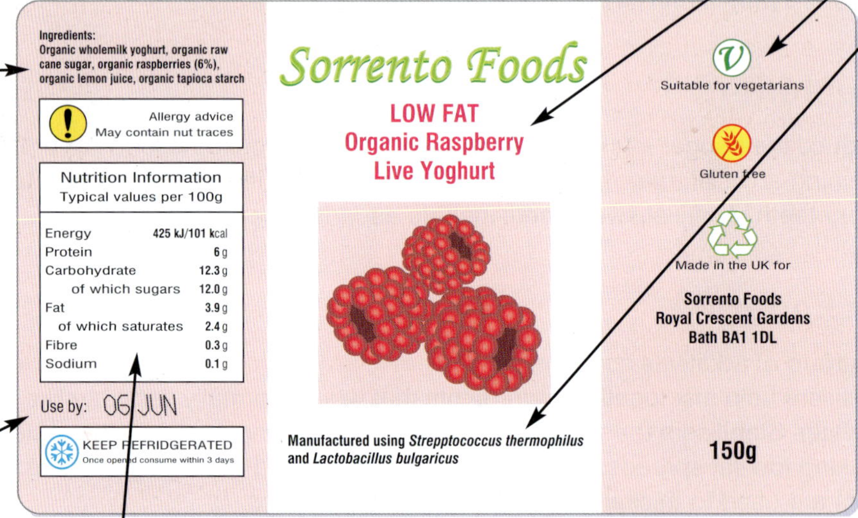

Ingredients

Ingredients are listed in descending order of weight: the ***higher*** up the list, the ***more*** of that food the product contains. If chemicals have been added, they must be listed on the label.

Dates

'Best before' refers to the ***time*** during which a food product is at its best. It refers to the quality of a product, rather than whether or not it is safe to eat. When the date runs out, the food will not usually be harmful, but may begin to lose its flavour and texture.

'Use by' dates refer to the ***safety*** of a substance you are eating. You must not eat or drink a product after this date even if it looks and smells fine, because you could be putting your health at risk.

'Sell by' or 'display until' dates are used by some shops to help with ***stock control***. They are instructions for shop staff, not shoppers.

Nutritional information

Values are usually quoted 'per 100 g'. This is so that different weights of foods can be compared. To work out how much of a food component you are eating, you must remember to convert these values to the actual weight of the food! The label will tell you:

- How much of the four main nutrient groups (energy, protein, fat and carbohydrate) the food contains.
- The amount of fibre and sodium contained in the food.

Most labels will also tell you:

- How much of the total carbohydrate is sugars (natural and added), the remainder is mostly starch. You should gain most of your energy from starchy foods.
- How much of the fat is saturated: this is the type of fat that can raise blood cholesterol levels.

Look at the label for Sorrento Foods and answer the following questions:

a) What is the last possible date you should eat this yoghurt?
b) How much saturated fat does a 100 g portion of this yoghurt contain?
c) How should you store this product?

Other information

Legislation requires food manufacturers to list any ingredients that people may be allergic or intolerant to, like nuts and gluten. It may also tell you how the product should be stored, whether the products are organic and whether or not the product is suitable for vegetarians.

GET IT RIGHT!

Many numerical questions could ask about food labels. Make sure you show your workings for any question – even if your final answer is wrong, marks are normally awarded for each step of the calculation. There is also often a mark available for writing the correct unit on an answer – so make sure you use units.

Low-fat products

Many people choose to buy low-fat products believing that these are healthier for them. At the moment there is no law stating what 'low fat' means. Therefore even though manufacturers make these claims, there may be very little difference between a 'low-fat' biscuit and the manufacturer's standard version. To check for yourself you need to compare the nutritional information sections of the labels.

A low-fat spread that could be used as an alternative to butter

d) A standard yoghurt contains about 15 g of fat per 100 g. Is Sorrento Foods' claim true that the yoghurt opposite is low fat?

GM free products

Some manufacturers are now labelling their products as 'GM free'. This is not an official symbol, but a manufacturer would be breaking the law if the claim was false. Genetically modified plants and animals are those which have had their genes altered. Scientists have used this technique to produce plants and animals with desirable features such as frost and pest resistance, and higher yields.

This manufacturer is trying to attract buyers by stating that its product contains no genetically modified products

e) Manufacturers must state if a food contains genetically modified organisms or their products. Are there any in Sorrento Foods' yoghurt opposite?

SUMMARY QUESTIONS

1 Copy and complete the following sentences using the words below:

nutritional **'use-by'** **ingredients**

Food labels contain a list of the . . . that the product contains. It also tells you when you must eat the product by, i.e. the . . . date, and information about the . . . content of the food.

2 Supermarkets often reduce food products on their 'best before' date. Can you buy these products and still be able to eat them the next day?

3 Summarise the information that can be gained from a food label.

4 Explain how the information contained in a food label could be used to help plan a healthy diet.

KEY POINTS

1 Food labels provide information about the ingredients a product contains, including any additives, advice on how you should store the product and its nutritional value.

2 A 'sell by' date provides shops with information for stock control. A 'best before' date refers to the quality of a food product, whereas a 'use by' date refers to the safety of a product.

3.9 Using microorganisms in food production

LEARNING OBJECTIVES

1 Which microorganisms can we use to produce foods?
2 How are bacteria used to produce yoghurt and cheese?
3 How do we use yeast to produce bread, beer and wine?

Bacteria and fungi play an important role in the production of some foods and drinks. Microbiologists study these microorganisms to find out the optimum conditions for their growth, and how this can be controlled to produce useful products.

Yeast is a type of fungus. It is needed to make bread, beer and wine. These three products are made using a chemical reaction called **fermentation**.

Fermentation is an example of **anaerobic respiration** – the yeast respires without oxygen to ferment sugar, producing alcohol and carbon dioxide. Fermentation can be summarised in this chemical equation:

glucose → ethanol + carbon dioxide
(alcohol)

$$C_6H_{12}O_6 \rightarrow 2C_2H_5OH + 2CO_2$$

a) What is ethanol more commonly known as?

Enzymes in yeast speed up the process of fermentation. The ideal conditions for fermentation are a good supply of glucose, with no oxygen present, at a temperature between 15 °C and 25 °C.

GET IT RIGHT!

Note how the examiners want you to write an answer. If the question says 'give the word equation for fermentation', write your answer in words – for example you need to write 'carbon dioxide', not 'CO_2'.

How do you make bread?

Bread is made using the fermentation reaction. A baker mixes together flour, water, sugar and yeast to make dough. The dough is then left in a warm place to rise. Enzymes in the yeast change the sugar into ethanol and carbon dioxide. As the gas is trapped inside the dough it makes it rise.

The dough is then baked. In the oven the ethanol evaporates. The bubbles of gas expand, making the bread rise further.

b) Why is dough not put in the fridge to rise?

This baker is kneading bread dough and shaping it into loaves

How do you make beer and wine?

Beer is made from malted barley grains (malt). These are mixed with warm water, which converts the starch in the barley into maltose (sugar). Hops are added for flavour and the liquid is boiled. Once cooled it runs into a vast container where yeast is added. When the oxygen in the container runs out, fermentation starts and continues until all the maltose is used up. The beer is then put into barrels or bottles.

Wine is made in a very similar way. Grapes are crushed and yeast added. Yeast changes the grape sugars into alcohol. When fermentation is complete the wine is bottled.

c) What are the sugar sources needed to produce beer and wine?

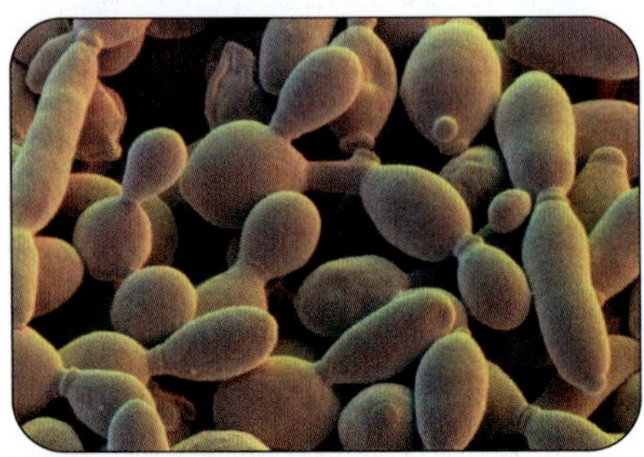

This is *Saccharomyces cerevisiae* – also know as baker's or brewer's yeast

Fermentation of milk

We use bacteria to ferment milk sugars in both cheese and yoghurt production. In both cases, bacteria convert lactose (milk sugar) into lactic acid. It is the acid that gives these products their characteristic tang.

How do you make cheese?

Cheese can be made from the milk of many animals. Bacteria are added to ferment sugar into lactic acid. Rennet is also added. This contains the enzyme rennin, which changes a milk protein into casein (curd). Milk curdles and separates into curds and whey. Whey (mainly water) is drained off and the curds are pressed to make the cheese solid. The cheese is left to ripen to improve its flavour and consistency.

d) Which two substances have to be added to milk to turn it into cheese?

How is yoghurt made?

Milk is boiled and bacteria are added. The milk is then kept warm for several hours. During this time, the bacteria multiply and ferment lactose into lactic acid. The lactic acid curdles the milk into yoghurt. It also restrains the growth of harmful bacteria, which increase the time that yoghurt can be kept and eaten safely. Yoghurts that have not been pasteurised to kill the bacteria used to make yoghurt are known as 'live' yoghurts.

e) What are the useful properties of lactic acid in yoghurt production?

This man is adding rennet to milk to make cheese. As well as using bacteria in cheese production, mould (a type of fungi) can also be used. For example, cheeses like Stilton and Danish Blue have mould growing through them.

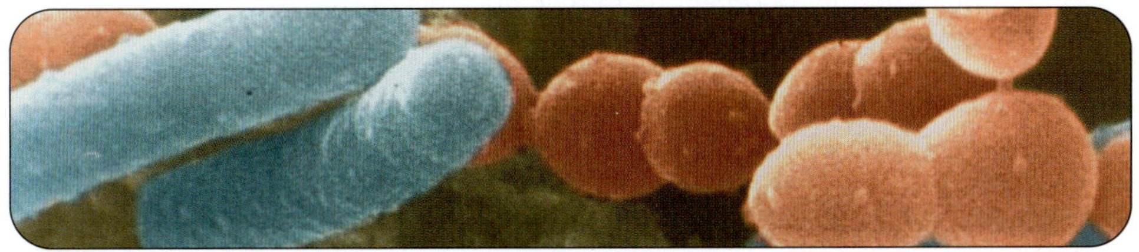

Streptococcus thermophilus (stained red) and *Lactobacillus bulgaricus* (stained blue) are two types of bacterium that are commonly used in yoghurt production

SUMMARY QUESTIONS

1 Copy and complete the following sentences using these words:

bread fermentation alcoholic lactose carbon dioxide curdles ferment yoghurt fungus

Yeast is a . . .. It is very important in the production of . . . and . . . drinks. During . . . , glucose is converted into ethanol and . . .. Bacteria are used to . . . milk in cheese and . . . production. The bacteria change . . . into lactic acid. This . . . the milk.

2 Why does bread not contain alcohol?

3 Summarise the production of beer or wine into a flow diagram.

4 Yoghurt and cheese are both produced using bacteria. Compare the similarities and differences between the ways they are manufactured.

KEY POINTS

1 Bacteria and fungi play an important role in the production of many foods and drinks.

2 Bacteria are used to ferment milk sugars in both cheese and yoghurt production.

3 Yeast is used to ferment sugars to make beer, wine and bread.

3.10 Food poisoning

LEARNING OBJECTIVES

1 What causes food poisoning?
2 What are the symptoms of food poisoning?
3 How can we prevent food poisoning?

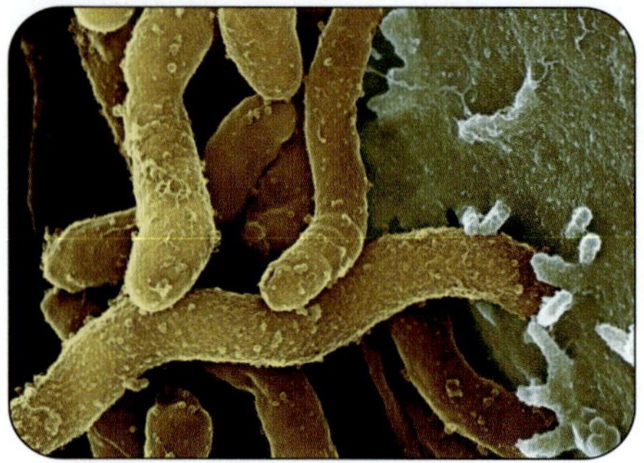

These are *Campylobacter* on the surface of the human stomach

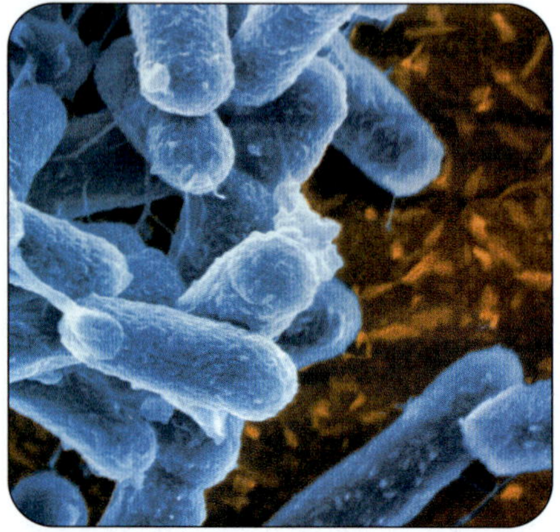

This is *E. coli* 0157. Bacteria multiply rapidly – in the right conditions one bacterium can multiply to more than 4 million in just 8 hours. To do so they need moisture, food, and warmth.

DID YOU KNOW?

There are more living organisms on the skin of a single human being than there are human beings on the surface of the Earth! Make sure you wash your hands thoroughly before cooking food.

Food poisoning is caused by the growth of microorganisms in food. The most serious types of food poisoning are due to bacteria and the toxins they produce. Hygiene and quality control staff, and Environmental Health Officers, are responsible for controlling the growth of bacteria in places such as restaurants and food manufacturing companies.

What types of bacteria cause food poisoning?

There are three main groups of bacterium that cause food poisoning:

- *Campylobacter* – This is the most common cause of food poisoning. *Campylobacter* can be found in raw meat, unpasteurised milk, and untreated water. Illness can be caused by a small number of bacteria.
- *Salmonella* – This is the second most common cause of food poisoning. Salmonella has been found in a wide range of food products including raw meat, eggs, raw unwashed vegetables and unpasteurised milk. It is found in the gut and faeces of animals and humans. Usually large numbers of this bacteria are needed to cause infection.
- *E. coli* – This organism is normally found in the guts of animals and humans. There are many different types, some of which are capable of causing illness. *E. coli* 0157 can cause severe illness and is found in raw and undercooked meats, unpasteurised milk and dairy products. A small number of these bacteria can lead to illness.

All these bacteria can survive refrigeration and freezer storage, but thorough cooking of food and pasteurisation will kill them.

a) What conditions do bacteria need to multiply?

What are the symptoms of food poisoning?

Common signs of food poisoning are stomach pains, diarrhoea, vomiting and fever. These symptoms usually appear rapidly, but can occur several days after eating contaminated food. If you suffer food poisoning you will normally get better within a few days. Over-the-counter medicines, such as oral re-hydration salts, can be taken to replace fluid and salts lost through diarrhoea. In more serious cases, fluids may need to be replaced through a drip. In rare circumstances food poisoning can kill.

b) How would you feel if you were suffering from food poisoning?

What procedures can be followed to prevent food poisoning?

Microorganisms are very hard to detect, as they do not usually affect the taste, appearance or smell of food. Microorganisms can enter our food at any time, from when an animal or crop is growing until the food is eaten. If they survive and multiply, they can cause illness.

There are a number of procedures we can use to ensure that foods are prepared with as little risk of contamination as possible. They include:

- Good personal hygiene practices, e.g. washing hands, tying back hair, covering cuts.
- Using detergents to ensure the working area remains clean.
- Using disinfectants to kill bacteria on work surfaces.
- The use of sterile packaging materials.
- Disposing of waste into appropriate containers, and regularly removing them from food preparation areas.
- Adequate control of pests, e.g. insects, mice.

c) Why should you disinfect kitchen worktops regularly?

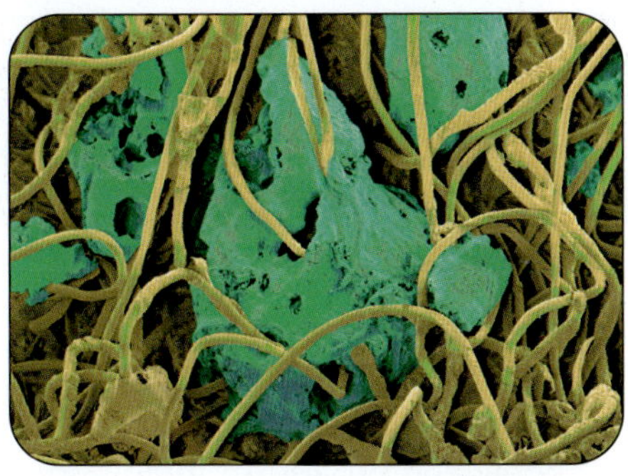

These are food particles trapped in a cleaning cloth. They provide an excellent breeding ground for bacteria. Hands, work surfaces and utensils must be cleaned regularly to prevent the cross-contamination of clean areas.

How are bacteria prevented from growing in food products?

The growth of bacteria can be slowed down or stopped altogether by changing the temperature. Bacteria multiply rapidly between about 5 °C and 65 °C. Most are killed at temperatures above 70 °C. Below 5 °C most bacteria multiply very slowly, if at all. At very low temperatures some bacteria will die, but many survive and will begin to multiply if conditions warm up.

Three other techniques can be used to preserve foods:

- ***Pickling*** – Acid (usually vinegar) is added to the food. Most bacteria cannot survive in highly acidic foods, e.g. pickled onions.
- ***Salting*** – Most bacteria cannot survive in very salty conditions. They will become dehydrated and die, or become temporarily inactivated, e.g. corned beef.
- ***Drying*** – Bacteria need water for growth. Without water their growth will slow and eventually stop, e.g. rice.

d) Why does cooking chicken thoroughly prevent you suffering from food poisoning?

For 'use by' and 'best before' dates to be accurate, you must follow storage instructions carefully e.g. 'keep in the fridge after opening'. If you don't, the food will spoil more quickly putting you at risk of food poisoning.

SUMMARY QUESTIONS

1 Copy and complete the table to show how bacteria in food can be prevented from growing:

Technique	Why are most bacteria prevented from growing?
Cooking	
Freezing	
Salting	
Pickling	
Drying	

2 Why does food keep for longer when stored in a fridge?

3 Why does vacuum packing or placing food in an airtight container help to prevent food poisoning?

4 Produce a list of good practices that could be used in a restaurant kitchen to ensure that food is produced as safely as possible.

DID YOU KNOW?

Each year it is estimated that as many as 5.5 million people in the UK suffer from food-borne illnesses – that's 1 in 10 people!

KEY POINTS

1 Food poisoning is caused by the growth of microorganisms in food.

2 Common signs of food poisoning are stomach pains, diarrhoea, vomiting and fever.

3 Good hygiene practices reduce the risk of food becoming contaminated with microorganisms. Thorough cooking and pasteurisation will kill any microorgansims present.

3.11 Growing crops

LEARNING OBJECTIVES

1 What is the difference between organic farming and intensive farming?
2 What chemicals do intensive farmers use to increase crop yields?
3 What alternative techniques do organic farmers use to increase crop production?

This farmer is spraying a ploughed field with chemical fertiliser

There are two different approaches to food production – **organic farming** and **intensive farming**:

- Intensive farming produces large quantities of food cheaply and efficiently by maximising the growth of crops and farm animals. Farmers use controlled environments and a number of different chemicals to achieve this.
- Organic farming uses natural methods of producing crops and rearing animals. Organic farmers only use a small number of artificial chemicals allowed by law. They let their animals roam as freely as possible. Many people believe food grown in this way is healthier and tastes better, and are willing to pay more for organic products.

People have different opinions on the ethical, economic and environmental impact of both farming techniques. Your views may also differ between crop production and the rearing of animals. (See pages 50–1.)

a) What type of products are normally more expensive to buy: organically or intensively farmed products?

Crop production

In order to produce a high yield of crops, farmers need to control four main factors: the nutrient content of the soil, and the presence of pests, weeds and fungi.

Adding nutrients to the soil

For healthy growth plants need the minerals nitrogen, phosphorus, potassium and magnesium. As crops grow they remove these essential nutrients from the soil. To continue to produce a high yield of crops from the same land, these nutrients need to be replaced:

- Intensive farmers use artificial chemical **fertilisers**. These fertilisers give plants the nutrients they need to grow effectively. The most widely used fertiliser in the UK is known as 'NPK fertiliser'. It contains nitrogen, phosphorus and potassium.
- Organic farmers add nutrients to the soil by adding manure or compost. Manure is better for the soil than artificial fertilisers. However, it can require the use of more heavy machinery. Alternatively they can rotate crops, as different crops take different nutrients from the soil. They can also plant leguminous plants like clover, because they add nitrates to the soil.

b) What do N, P and K stand for in chemical fertilisers?

Dealing with pests

- Many insects eat crops, e.g. aphids, locusts and beetles. Intensive farmers deal with this problem by spraying crops with chemical pesticides that kill the insects.
- All crop pests have natural predators. Farmers can exploit this relationship to kill pests. This is called **biological control**. Predators (normally other insects) are bred in large numbers. They are then released onto the crops where they eat the pests.
- Sometimes moulds and fungi are used to kill the pests. This works by the microorganism infecting the pest with a disease.
- Farmers can also use selective breeding techniques. These can produce new varieties of crops that are more resistant to pests and disease.

c) Name an example of an organism that can be used to control aphids.

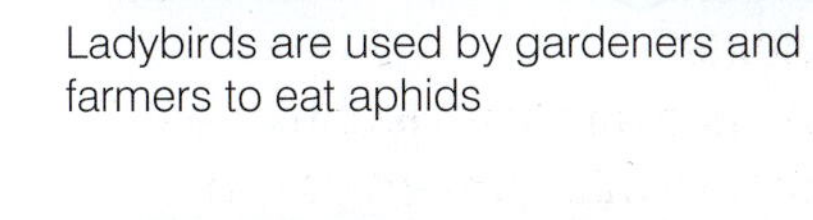

Ladybirds are used by gardeners and farmers to eat aphids

Dealing with weeds

- **Herbicides** are chemicals widely used by intensive farmers. Herbicides kill weeds (other plants) that would compete with the crop for water, nutrients and space.
- On a small scale, organic farmers can remove weeds by hand. Machines have been developed to help weed large crop areas without damaging the crop. This method works well on crops that are grown in rows, such as vegetables.

This man is weeding his vegetable plot. This is an extremely time consuming process.

Dealing with fungi

- **Fungicides** are other chemicals used by intensive farmers. These chemicals kill fungi that can damage large areas of crops.
- Organic farmers rely on growing strong healthy crops to combat disease. However, if a microorganism does attack an organic crop, the farmer will remove the infected material and dispose of it, normally by burning.

d) What is the difference between a herbicide and a fungicide?

SUMMARY QUESTIONS

1 Copy and complete the following summary table to describe how farmers treat potential crop growing problems:

Farming problem	Intensive farmers	Organic farmers
Pests		
Weeds		
Fungi		

2 Describe three methods organic farmers use to ensure their crops receive an adequate supply of nutrients.

3 Why is it particularly important to wash intensively farmed fruit and vegetables before eating them?

4 Produce a table comparing the advantages and disadvantages of producing crops intensively and organically.

KEY POINTS

1 Intensive farming relies on the use of a range of chemicals to produce crops with high yields. Organic farming generally produces less crops from the same area of land. Many people think organic food tastes better and is healthier.

2 Common chemicals used by intensive farmers are pesticides, herbicides, fungicides and fertilisers.

3 Organic farmers add nutrients to soils using manure and compost. They kill pests using biological control and remove weeds by hand or using a machine.

3.12 Rearing animals

LEARNING OBJECTIVES

1 How are animals reared intensively to produce cheaper products?
2 How are animals reared organically?

Although most people are happy to buy intensively farmed crops, some are less happy to buy intensively reared animals. Intensively farmed animals are kept in a strictly controlled environment which makes the animals increase in size quickly. This makes intensively farmed animals and their products cheaper.

However, some people raise concerns over the animals' well-being. They believe the conditions they are kept in are unethical.

The table below summarises the main differences between how animals are reared using intensive and organic methods:

Factors that can be controlled	Intensively reared animals	Organically reared animals
Food supply	Animals are fed a high protein diet to rapidly increase their body mass.	Organic food is fed to the animals, e.g. organically grown hay, grass and silage are often used as a food supply for cattle.
Temperature	Animals are kept indoors, in a warm environment. Animals waste less energy heating their own bodies.	Animals would normally live outdoors. In bad weather animals may be kept inside.
Space	Restricted movement. Animals do not waste energy moving around.	Animals are allowed to roam as freely as possible.
Use of drugs	Antibiotics are often given to animals to prevent the spread of disease.	Antibiotics are not used unless an animal is ill.
Safety of enclosure	Animals are kept safe from predators.	Animals are often kept indoors at night to protect them from predators.
Infection control	Animals are close-packed, so infections spread quickly if they get in. The animals are checked regularly and the sheds are under strict quarantine.	Animals kept outside are more likely to catch infections, such as bird flu, from wild birds.
Waste efficiency	Animal waste can be collected and converted into biogas.	Animal waste is not available for biogas.

GET IT RIGHT!

If you are asked to compare the two types of farming, make sure you stick to the facts. Answers such as 'intensive farming is cruel' will not gain credit. To gain a mark you could say 'some people think intensive farming is cruel as animals are kept in very small spaces'.

a) Why do intensive farmers give antibiotics to healthy animals?
b) How does the amount of space an animal is provided with differ between intensive and organic farming?

Battery-farmed versus free-range egg production

Around two-thirds of the total UK egg-laying hen population are kept in battery cages. On average, these hens will each lay 300 eggs a year. This is over 10 times the number produced by a wild hen!

The cages are stacked in large windowless sheds which are kept pleasantly warm. These sheds often accommodate more than 20 000 birds. A typical cage contains four or five birds. The minimum space allowed per bird is 550 cm^2. This is less than the surface area of a standard sheet of A4 paper.

The hens have an automated food and water supply. Light conditions inside the shed can be altered so that they depict spring, with long daylight hours. This is when hens naturally lay more eggs. Therefore egg production is increased.

The cage system is designed to allow faeces to drop through the bottom of the cage. This separates the hens from possible sources of disease. However, these cages cause suffering for the hens that live in them. Battery hens often suffer from foot deformities caused by the absence of suitable perches and restrictions on movement.

The birds are also prone to multiple fractures, caused by bone weakness. This is due to the high rate of egg production, resulting in calcium deficiency.

Battery-farmed hens are kept in a carefully controlled environment, which significantly increases egg production but unfortunately reduces the hen's life expectancy

In comparison, organic farmers must provide at least an acre of field for every 400 chickens. The hens are free to roam over pasture during daylight hours, but may choose to shelter indoors in cold, wet or windy weather. During these periods, space is provided for the hens to move around and perch.

Specially developed nest boxes give the birds the quiet and security they need to lay. They also ensure that the eggs can be collected quickly.

These hens are roaming free on a farm. They will produce fewer eggs than battery-farmed hens

c) Explain why hens kept in battery conditions can suffer from broken bones.

SUMMARY QUESTIONS

1 Copy and complete the following table using the most appropriate words:

Factor	Battery hens	Free-range hens
Space available	Large / Small	Large / Small
Access to outdoor space	Yes / No	Yes / No
Life expectancy	High / Low	High / Low
Number of eggs produced	More / Less	More / Less
Cost of eggs	High / Low	High / Low

2 Why do you think free-range eggs cost up to twice as much as battery-farmed eggs?

3 Explain the factors that an intensive farmer could control to increase the rate of meat production in pigs.

KEY POINTS

1. Intensively farmed animals are kept in a strictly controlled environment which makes the animals increase in size quickly. Factors that are controlled include their diet, size and temperature of enclosure and the routine use of drugs.
2. Organically farmed animals are allowed to live a more natural life, where they are free to roam in large enclosures and only given drugs when ill.

3.13 The impact of intensive farming on the environment and animals

ACTIVITY

1 In pairs, make a list of all the damaging effects intensive farming can have on the environment.

Destroying our environment?

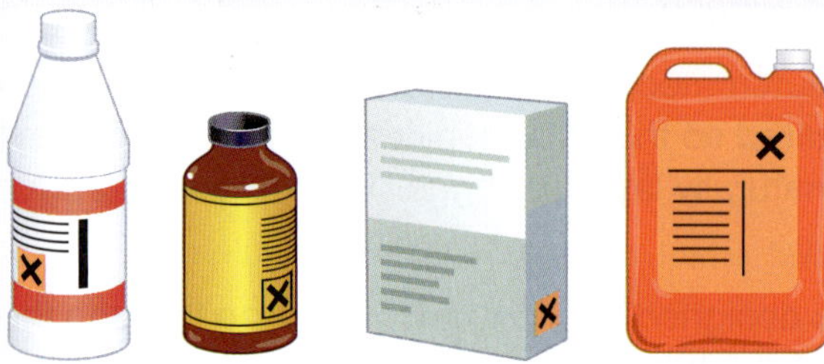

Intensive farming produces as much food as possible, by making the best use of land, plants and animals. Intensive farmers maximise their output by various methods – specifically breeding animals for high growth rates, using controlled environments and a number of different chemicals. This makes the food produced as cheap as possible. However, this highly intensive farming can have huge environmental impacts and affect an animal's well being.

This algal bloom was photographed in marshland by the Thames Estuary. It is caused by fertiliser running off the land into the water

ACTIVITY

2 Produce a cartoon to explain how eutrophication occurs – this could be given to farmers to highlight the potential problems caused by fertilisers.

Polluting our rivers?

When farmers apply a large amount of fertiliser to their land, lots may be wasted. When it rains, it drains away from the soil into ponds, lakes and rivers. The fertiliser in the water allows algae to grow rapidly, quickly covering the surface. This stops light reaching the lower plants. These dying plants are broken down by bacteria in the water. The decaying process uses up lots of oxygen from the water, making it difficult for animals to survive and many fish die. This is called **eutrophication**.

If high levels of nitrates (present in fertiliser) enter our drinking water, it can damage our health. Bacteria change the nitrates to nitrites. These can stick to red blood cells, stopping them carrying oxygen.

ACTIVITY

3 Discuss the preventative steps farmers could take to prevent soil erosion affecting an area of land.

Eroding our inheritance?

Over-grazing of land leaves the area exposed to ***soil erosion***. This is particularly a problem on hillsides, which are fully exposed to the strength of the winds and the rain as it runs down the hill. Over-grazing results in the loss of protective vegetation – mainly grass. Grass roots play an important role in binding soil together helping it to maintain its structure and coherence. Barren soil dries out easily, and becomes increasingly susceptible to erosion by water and wind.

This barren hillside has been severely affected by soil erosion, as a result of the over-grazing of farm animals

Quality of life?

Pigs are often reared in a pig battery, in similar conditions to battery hens. This is a highly controlled environment. They have restricted movement and are fed a high protein diet to speed up their growth. Many pigs are huddled together in a small space, keeping each other warm and protecting them from predators.

Pigs are kept in cramped conditions in pig batteries

Energy costs money, so farmers try to waste as little energy as possible from the animal itself or the environment in which it is kept. As a result of being kept in this controlled environment, the pigs are less able to perform natural acts, e.g. scratching or grubbing in soil, and rolling in dirt.

Free range pigs are more aggressive towards each other. Kept in fields, they can get sunburnt and suffer heat exhaustion in the summer. These pigs are also more likely to pick up infections.

ACTIVITY

4 Meat produced intensively is much cheaper. Would you be prepared to pay more for organically produced bacon? Make a list of the reasons that made you reach your decision.

ACTIVITY

5 Research the type of animal that can be found living in a hedgerow.

Destroying habitats?

Many intensive farmers remove hedgerows to create larger fields. This creates more space for growing crops and the land becomes easier to farm as large machinery can be used. However, hedges are the homes for hundreds of plants and animals; removing them destroys many organisms' habitats. It also causes soil erosion, as the hedges act as natural wind blocks. Once hedges are removed, when the fields are barren during winter soil can be blown or washed away.

A natural hedgerow

Poisoning wildlife?

DDT is a very effective pesticide – only very small amounts are needed to kill an insect. DDT was a widely used pesticide which saved many people from starvation by killing crop pests. It has also been used to kill the mosquitoes that spread malaria. However, it has killed large numbers of wildlife, as it is toxic at high concentrations. DDT does not decompose easily. Because of this, it passes along a food chain until it reaches fatal levels – killing many top predators. As a result the UK (and many other countries) ban the use of DDT.

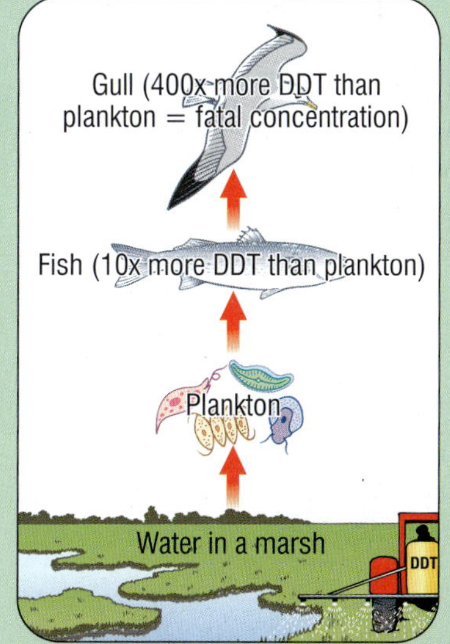

DDT accumulation in a food chain

ACTIVITY

6 Design an A4 poster to be displayed in garden centres to make people aware of potential dangers of spraying plants with insecticides in their gardens.

SUMMARY QUESTIONS

1 Complete the sentences using the words below:

calcium carbohydrates deficiency diet disease fats fibre minerals nutrients proteins vitamins

The human body requires a variety of ...... to carry out the vital life processes. These include ......, ... and ......, ... is also needed for healthy bowel movements, ... and ...... are needed in small amounts to remain healthy. For example, ... is needed for strong teeth and bones. If you do not take enough of a vitamin into your body through your ......, you will develop a vitamin .... If this occurs over a period of time it can cause ....

2 Complete the sentences using the words below:

additives behavioural blood chemicals diabetes energy flavour heart ingredients saturated shelf life sodium store sugar tired value weight

You should limit your intake of ...... fat as this can cause ...... disease. Too much ...... in your diet can cause high ...... pressure and too much ...... can lead to ....

It is important to control your ...... intake. Too little and you will be very ...... and lethargic. Too much and you will put on ....

Additives are ...... added to food to enhance its ......, appearance and .... Some additives have been linked to ...... problems in children.

All food and drink products have a food label. These provide you with information about the ...... the food contains, including any ......, advice on how you should ...... the product and the nutritional ...... it gives you.

3 Complete the sentences using the words below:

bacteria beer bread cheese chemicals clean cooking harmful intensive microorganisms natural organic poisoning storing yeast yields yoghurt

.... are used in the manufacture of many foods and drinks. ...... are involved in making ...... and ......, and ...... is used to make ...... and ....

The growth of ...... bacteria and fungi in food can cause food .... This can be prevented by keeping kitchens......, ... food thoroughly and ...... food correctly.

There are two approaches to farming. ...... farming relies heavily on ...... to maximise crop and animal .... farming is more ......, no artificial chemicals can be used.

KEY PRACTICAL QUESTIONS

1 Describe how to:

a) Test for the presence of saturated fat in a food.

b) Test for the presence of glucose in a food.

c) Sterilise a wire loop.

d) Carry out a titration to test the acidity of a food.

e) Make a streak plate.

2 Why do we use these?

a) DCPIP

b) Biuret solution.

c) Iodine.

d) Aseptic technique when working with microorganisms.

e) Serial dilutions to count the number of bacteria in a sample.

EXAM-STYLE QUESTIONS

1 a) Microorganisms play an important role in the production of a number of food and drink products. Match the microorganism to the product it is involved in making: (2)

	ham
yeast	cheese
bacteria	wine
	jam

b) To make beer, brewers add yeast to the malt. What process does the yeast carry out to turn this into alcohol? (1)

c) As well as producing alcohol, what gas is also made by this process? (1)

d) Describe how bread is made from flour, water and sugar. (3)

2 The label below comes from a container of ready-made custard.

Mother's Own Custard

1 kg

Shake well before opening

Once opened keep refrigerated and eat within 3 days

Ingredients: milk, sugar, starch, flavourings, colour

Contains no artificial colours or preservatives

Nutritional information	
Typical values of nutrients	per 100g of product
Energy	420kJ
Protein	2.8g
Carbohydrate	15.6g
Fat	3.1g
Fibre	0.0g
Sodium	0.2g

a) The label gives the mass of nutrients in 100 g of product:

i) Calculate the mass of nutrients in 100 g of product. (2)

ii) Name one substance that is not listed in the ingredients, but which must be present in 100 g of product. (1)

iii) Calculate the total mass of protein in the container of custard. (2)

b) The label contains some important advice about storing the product:

i) Why should the custard be kept refrigerated after opening? (1)

ii) Why must the custard be eaten within 3 days of opening? (1 mark)

c) The custard contains no artificial colour or preservative:

i) What is a preservative? (1)

ii) Ready-made custard used to contain artificial colour and preservative. Give two reasons why this custard is now sold with no artificial colour or preservative. (2)

3 Most of the food grown in the UK today is produced by intensive farming methods. These methods give us large volumes of high quality food at low cost. In contrast, the volume of food produced by organic farming methods is small.

a) Explain the alternative methods used by organic farmers to produce good crop yields without the use of artificial fertilisers, pesticides, herbicides and fungicides. (6)

b) The demand for food produced by organic farming methods is increasing, despite the fact that this food is generally more expensive. Give two reasons why the demand for food produced by organic farming methods is increasing. (2)

4

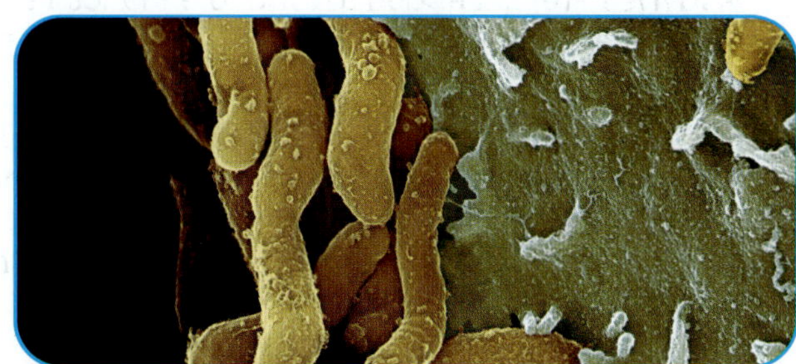

a) What causes food poisoning? (1)

b) Name a specific type of the microorganism shown in the photo that is a common cause of food poisoning. (1)

c) Describe the common symptoms of mild food poisoning. (2)

d) Name two practices you should follow at home to prevent food from being contaminated with microorganisms. (2)

e) To prevent bacteria growing in food products, manufacturers either add substances to the food, or treat the food in some way. Name two of these techniques and explain how they work. (4)

EXAMINATION STYLE QUESTIONS

1 Archie is trying to lose weight, and has been referred to a dietician by his doctor. The dietician asked Archie to bring with him a diary of all the food and drink products he has eaten in the last week. Using the extract from his diary answer the questions below:

> **Tuesday lunch time**
>
> Main meal – Roast chicken, deep fried chips, peas and carrots
> Pudding – Apple crumble and cream
> Drink – Orange juice and lemonade

(a) Choose one type of food Archie has eaten which contains each of the following nutrients:

Nutrient	Food/Drink
Carbohydrate	
Protein	
Fat	
Vitamins	

(*4 marks*)

(b) What is protein used for in the body? (*1 mark*)

(c) Name a component of Archie's diet that is rich in the mineral calcium. (*1 mark*)

(d) Which mineral is needed by the body for making red blood cells ? (*1 mark*)

(e) (i) Archie cooked his chips in olive oil. Is this a source of saturated or unsaturated fat? (*1 mark*)

(ii) Why is eating lots of saturated fat bad for your health? (*1 mark*)

(f) Archie added salt to his chips. What is the health risk of eating lots of salt? (*1 mark*)

2 Dylan is an intensive farmer who raises his 2000 hens on a small battery farm. Emily is trying to convince him that rearing hens organically is better for the chickens he keeps, and will not reduce his profits. To help Dylan reach a decision, Emily provides him with the table below, comparing the two types of farming. Look at the data in the table.

Factor	Battery hens	Organic hens
Space available	Large shed	5 acres
Access to outdoor space	No	Yes
Life expectancy	About 2 years	About 5 years
Number of eggs produced per hen	Around 240 a year	Around 80 a year
Cost of eggs	45p for 6	90p for 6

(a) State 2 advantages for Dylan of rearing his hens organically. (*2 marks*)

GET IT RIGHT!

Part (a) asks you to choose foods from the list provided. Make sure you select your answers from the information given. Any other answer (even if correct) will not gain credit. Read the question carefully!

(b) Dylan is concerned that changing his farming practice would cost him money. He knows that his battery hens produce around 480 000 eggs per year, earning him £36 000 per year.

(i) Calculate how many eggs would be produced per year if he farmed the same number of hens organically. *(1 mark)*

(ii) Calculate the income Dylan would earn in a year from farming his hens organically. How does this compare with the income he currently earns from farming his hens intensively *(3 marks)*

(iii) Name one way he would save money by rearing hens organically. *(1 mark)*

(c) Emily suggests that as well as raising his hens organically, Dylan should grow his own feed for the hens. He already spreads his fields with manure, but is unsure how he could manage without using pesticides and herbicides. State and explain how these problems can be dealt with without using chemicals. *(3 marks)*

3 (a) Microbiologists detect and analyse the growth of microorgansims in food products. The presence of some microorganisms in food products can be damaging to health, but others are used to produce specific food and drink products.

(i) Yeast is used in the production of bread. What type of microorganism is yeast? *(1 mark)*

(ii) What useful chemical reaction does yeast perform in the production of bread? *(1 mark)*

(iii) This chemical reaction can be summarised in the following word equation:

glucose → ethanol + carbon dioxide

$$\ldots\ldots\ldots\ldots \rightarrow 2C_2H_5OH + \ldots\ldots\ldots\ldots$$

Complete this chemical reaction using the correct chemical formulae. *(2 marks)*

(b) Yoghurt is an example of a food product made using bacteria. Milk is boiled and bacteria are added. The milk is then kept warm for several hours. During this time the bacteria multiply and ferment lactose into lactic acid.

(i) State the two reasons why lactic acid is important in this process. *(2 marks)*

(ii) Some yoghurts and other products made from milk are pasteurised. Why can you store pasteurised food safely for a longer period of time than non-pasteurised food? *(1 mark)*

(c) Many people do not like chemical additives being added to their foods, as there are potential health risks associated with the use of some of these chemicals.

(i) What is the name of the group of chemical additives that can be added to foods to increase their 'shelf-life'? *(1 mark)*

(ii) Choose a different group of chemical additives. Explain why they are used in the food processing industry, and state a possible health risk associated with their use. *(3 marks)*

GET IT RIGHT!

This question involves lots of calculations. To ensure you gain full credit for your work, make sure you set out your working clearly, so that the examiner can follow what you are doing. You also need to make your answer clear – it is a good idea to underline this. Questions which state 'Show your working' mean just that! If you do not, you will lose marks.

GET IT RIGHT!

After each question it tells you how many marks are awarded. For each mark available you need to state one point. Part c(ii) is worth 3 marks, and therefore you have to make three points. This question makes it quite clear what these 3 points need to be: the name of a group of chemical additives, their use and a possible health risk associated with their use. Not all questions are this prescriptive; they may just say 'describe' or 'explain'. Writing your answer in bullet points is a good way of making sure that you state enough points: each bullet point should be worth one mark.

3.14 Food science coursework tasks

LEARNING OBJECTIVES

1 What is involved in the food science coursework investigation task?
2 How do I maximise my marks?

Food tasting research

Today food scientists study foods and food components, either for research purposes or to support our manufacturing industry. They are involved in:

- Analysing the quality and safety of food products.
- Investigating new manufacturing methods.
- Creating new food products.

Discuss a recent example of a food safety scare.

Often food scientists:

- Apply scientific methods to keep food fresh, safe and attractive.
- Research ways to produce food more quickly and cheaply.
- Check food quality, for example:
  - The amount of carbon dioxide gas or vitamins in a fizzy drink.
  - Physical properties, like the density of food.
- Work in ***'quality assurance'*** and food safety, possibly:
  - Checking types and numbers of microorganisms.
  - Developing food supplements and their effect on our health.

Remember, for your coursework, you need to produce a report of one practical investigation set in a vocational context. See pages 156–7 for the Coursework Checklist and mark scheme. 40% of your GCSE marks come from this one report. However, you should carry out more than one experiment as this will help you to perform better in the Unit 2 exam, and progress through the Unit 3 stages of this course.

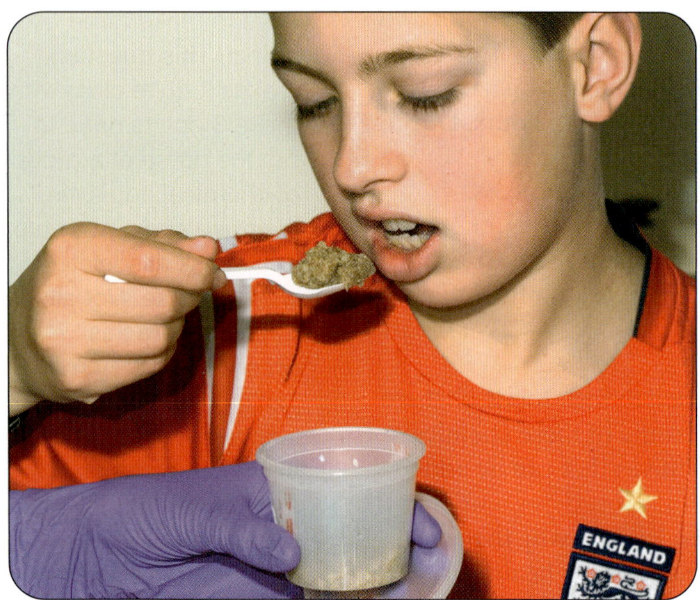

Food challenge. The boy eats a variety of foods. The foods include suspected ***allergens*** and ***placebos***. Some may cause a reaction and confirm he has a food allergy.

Food research into wheat grains. Wheat contains gluten. Gluten can damage people's intestines if they have the rare ***coeliac*** disease.

The food science investigation will relate to food or components that may be found in food or food supplements. In the report of your food science investigation:

- Describe the purpose of your investigation.
- Include a plan and risk assessment for your investigation.
- Make conclusions from, and evaluate, your investigation.
- Explain how a food scientist might use the results of your investigation.

The three food science investigation tasks are:

- The analysis of a fruit drink (pages 62–3): detecting starch, fat, glucose, sucrose and protein molecules, and measuring the vitamin C content, moisture content, suspended matter and acidity.
- Investigating bacterial growth (pages 64–5): the effect of different conditions on the growth of bacteria.
- Investigating the effect of nutrients on plant growth (pages 66–7): the effect of nutrients on the growth of a food product.

The following task is not detailed enough for a coursework investigation, but could be used as a mock investigation to develop your practical skills:

- Testing food supplements for iron content (pages 60–1).

Discuss whether or not you are happy with artificial ingredients being added to natural foods. Think of the advantages and disadvantages before you make your decision.

The domination of UK supermarket chains unduly affects both farmers and the public – the food they supply and the food we consume. Hold a debate within your class to explore your views.

SUMMARY QUESTIONS

1 Write down two examples of food supplements.

2 Think about 'quality assurance'. Give an example of how scientific work ***assures*** the public of the ***quality*** of a food product.

3 Read the caption under the photo of the food challenge opposite.

a) What is an 'allergen'?

b) A 'placebo' is an inactive substance used as a control in an experiment. Explain the point of giving the boy a food that contains no active chemical: 'Surely it could deceive him – giving the boy a food that does not harm him?

3.15 Testing food supplements for iron content

LEARNING OBJECTIVE

1 How do you test food supplements for iron content?

COURSEWORK HINTS

This task is designed to help you to improve your practical skills before you embark on a coursework task.

Iron is an essential mineral that the body requires in small amounts. It is widely found in foods including meat, dried fruits and green leafy vegetables. Iron from plant sources is absorbed only half as well as that from animal sources. The average diet provides about 10 to 20 mg of iron per day – the Recommended Daily Allowance (RDA) is 14 mg.

Why is iron needed by the body?

Iron is needed to manufacture haemoglobin, which enables red blood cells to transfer oxygen to the body's tissues. Severe or prolonged iron deficiency leads to anaemia.

Why are iron supplements prescribed for some people?

People who suffer from anaemia cannot manufacture haemoglobin effectively. Pregnant women require extra iron too, as they have up to twice the normal blood volume towards the end of pregnancy to support the growing baby, and to cover blood losses during delivery. Iron supplements are often prescribed in both cases to ensure that oxygen is carried effectively around the body.

Iron tablets – these may be prescribed by doctors to patients suffering from anaemia

Testing the iron content of food supplements

It is important that people are aware of the precise amount of iron taken, as too much can be toxic, and too little would not alleviate the symptoms from which they are suffering. Scientists, working in quality control, regularly sample products to ensure that their contents exactly match the description of the product.

a) Why is it important to ensure that the amount of a vitamin or mineral supplement is accurately listed on a label?

PRACTICAL

Finding the mass of iron in a tablet

The procedure used to test an iron tablet for its iron content is as follows (wear eye protection throughout the whole procedure):

1 Grind up five iron tablets into a fine powder.
2 Add 25 cm^3 of 1 mol/dm^3 sulfuric acid (H_2SO_4) and stir until the tablets dissolve. Warm gently using a water bath if necessary. Be careful, this is an irritant.
3 Transfer the contents to a 250 cm^3 volumetric flask.
4 Wash all apparatus used with distilled water. Transfer these washings to the volumetric flask, and top up to the 250 cm^3 mark with 1 mol/dm^3 sulfuric acid.

5 Using a pipette, transfer 25 cm³ of this solution to a conical flask.
6 Fill a burette with 0.01 mol/dm³ potassium manganate(VII) solution (purple in colour). Drain some of the liquid into a beaker to remove air from below the tap.
7 Record the starting volume to the nearest 0.1 cm³.
8 Add the potassium manganate(VII) solution ($KMnO_4$) to the conical flask, swirling the flask as you do this. The purple solution will be decolourised.
9 At the first sign of a pale pink colour in the conical flask solution, stop the titration. Record the volume on the burette.
10 Calculate the volume of potassium manganate(VII) solution added. This is your rough titration.
11 Repeat the titration more carefully, swirling and adding the potassium manganate(VII) solution slowly, drop by drop, near the end point. Continue repeats until you get two accurate volumes within 0.1 cm³. Average the two accurate titrations.
12 Use the graph below to calculate the mass of iron (per tablet) in the tablets that you tested.

A mortar and pestle are often used to grind a solid material down to a fine powder

b) An investigation was carried out using the above procedure to calculate the iron content of a mineral supplement tablet. An average of 25 cm³ of potassium manganate(VII) solution was required to produce a pale pink solution. What was the mass of iron in each of the tablets tested?

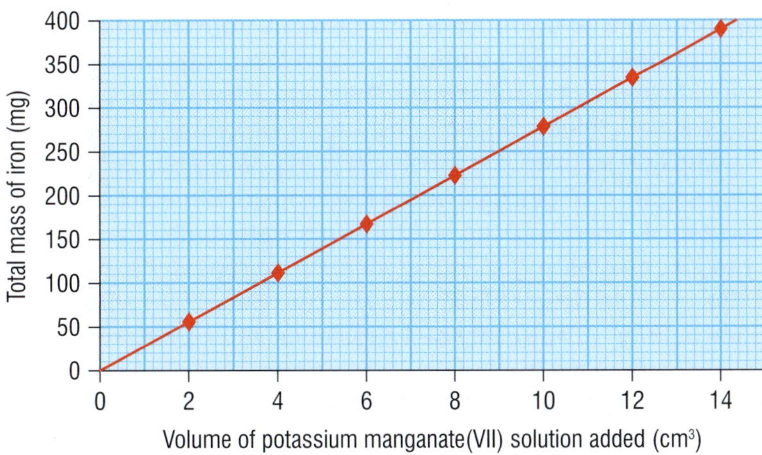

The mass of iron tablets

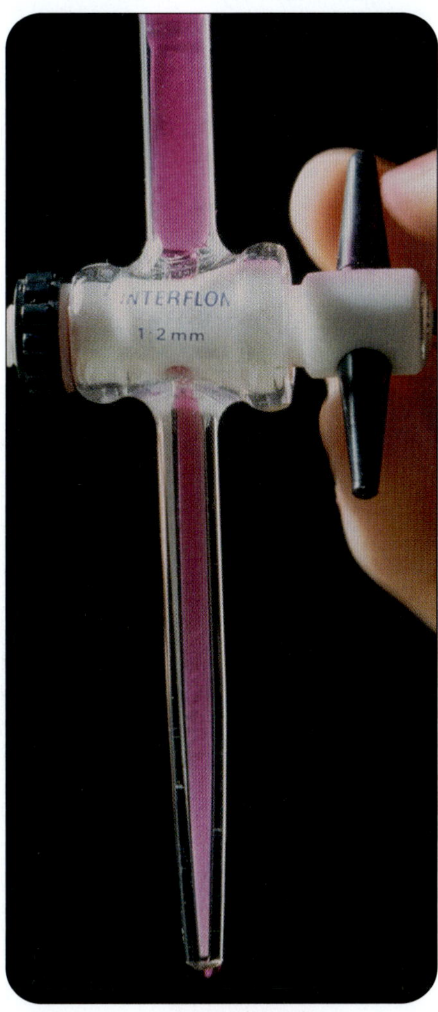

Removing the air from below the tap of a burette

COURSEWORK TASKS

You are going to carry out a comparison into the iron content of a number of different iron tablets. You will need to be able to carry out the titration on these pages. Before beginning this task carry out a risk assessment and practise the technique.

3.16 The analysis of a fruit drink

LEARNING OBJECTIVES

1 How do you test foods ***qualitatively*** for starch, fat, protein, sugar and acidity?

2 How do you tests foods ***quantitatively*** for moisture content, suspended matter, vitamin C content and acidity?

A positive result for a starch test

A positive result for a fat test

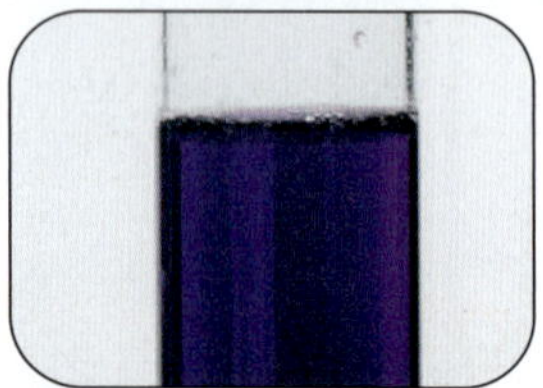

A positive result for protein

A positive result for sugar

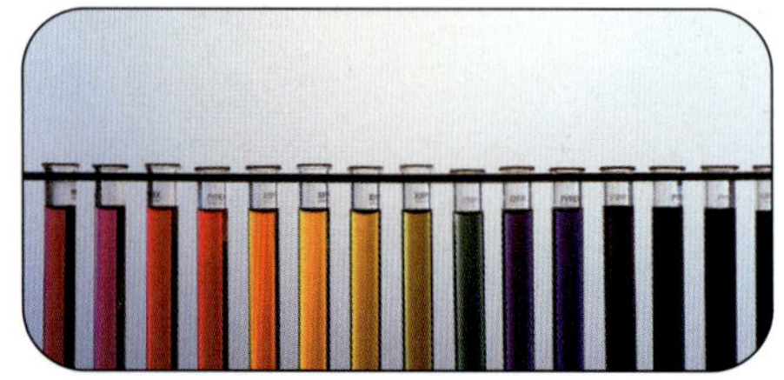

Different pH values produce different colours with universal indicator

Food scientists carry out a number of tests to determine what is actually in the foods and drinks that we are consuming. Some of these detect the presence of a substance – this is called a **qualitative test**. It tells you whether or not the food or drink contains a substance, but not how much is present. To find out how much is present, you need to carry out a **quantitative test**.

a) What is the difference between a qualitative and quantitative test?

Common qualitative tests performed by food scientists

For most food tests, scientists have to make a solution of the food. You can do this by crushing the food in a pestle and mortar. Add a spatula-full to a boiling tube, add 5 cm^3 of distilled water and stir. Whilst wearing eye protection, bring to the boil and simmer for one minute, then allow to cool before carrying out the appropriate test. (This is not necessary for the fat test.) For all the tests on these pages wear eye protection and avoid chemicals contacting the skin.

- **Testing for starch** – Add a few drops of iodine solution (orange-coloured) to the substance. If the solution turns a dark blue-black colour, then starch is present.
- **Testing for fat** – Add a few drops of ethanol to the food/drink, shake the test tube and leave for one minute. Pour the ethanol into a test tube of water. If the solution turns cloudy, it contains fat. (**Warning:** Ethanol is highly flammable so make sure you are a safe distance from naked flames.)
- **Testing for protein (Biuret solution)** – Add a few drops of copper sulfate solution (pale blue) to the substance. Then add a few drops of sodium hydroxide solution. If the solution turns purple, protein is present. (**Warning:** Sodium hydroxide solution is corrosive and copper sulfate is harmful.)
- **Testing for sugar** – For glucose, add a few drops of Benedict's solution (blue) to the substance. Then heat the test tube in a water bath. If the solution turns orangey-red it contains sugar. To test for sucrose repeat the above, but add a few drops of dilute hydrochloric acid after the Benedict's solution.
- **Testing for acidity** – The acidity of a substance can be determined using universal indicator in the form of liquid or a paper strip. The colour needs to be read in conjunction with a pH scale to determine the pH of the food being tested.

Common quantitative tests performed by food scientists

- **Moisture content by evaporation** – This test is carried out to find out how much liquid is in a food product. The food being tested is placed into a beaker, and the mass taken on a balance (the mass of the empty beaker must be subtracted). It is then placed in an oven overnight at 40 °C, and the mass taken again. The moisture content (% by mass) of the food can be calculated from:

$$\textbf{Moisture content (\%)} = \frac{\textbf{mass before} - \textbf{mass after}}{\textbf{mass before}} \times \textbf{100}$$

- **Suspended matter by filtration** – This test is carried out on liquid foods to find out how much undissolved solid material is present in the liquid. The total mass of the liquid food is taken before the test. The mixture is passed through a dried filter paper.

The filter paper and residue are then dried, and the mass of the residue measured. The suspended matter (% by mass) of the liquid food can be calculated from:

$$\textbf{Suspended matter (\%)} = \frac{\textbf{mass of residue}}{\textbf{original mass of liquid food}} \times \textbf{100}$$

b) A soup contained 45 g of suspended matter from a 200 g sample. What percentage of the soup was solid matter?

- **Vitamin C content**
  The procedure used is as follows:
  - If the food being tested is solid, dissolve a fixed amount of the food in acid to make a food solution. Liquid foods and drinks (e.g. fruit juices) do not need to be dissolved in acid first.
  - Place the food solution in a burette.
  - Place a fixed volume of DCPIP indicator in a conical flask. DCPIP indicator is a blue solution. Place the conical flask on a white tile.
  - Titrate the food sample into the indicator until the indicator becomes decolourised.
  - The more food solution that is required to decolourise the indicator, the less vitamin C is in the food.

c) Why is it important to use a fixed volume of DCPIP indicator if you are comparing the vitamin C content between different foods?

- **Acidity of a product by titration** – This test is carried out on liquids to test how acidic they are. 25 cm^3 of 0.1 mol/dm^3 sodium hydroxide (irritant) is added to a conical flask using a pipette, along with a couple of drops of phenolphthalein indicator (pink in alkaline solution). The product being tested is added to a burette, and titrated into the sodium hydroxide. At the point where the indicator goes colourless (the end point) the acidity can be calculated from:

$$\textbf{Concentration of acid (mol/dm}^3\textbf{)} = \frac{\textbf{conc. of alkali} \times \textbf{volume of alkali}}{\textbf{2} \times \textbf{volume of acid used}}$$

(**Note:** This equation applies only to citrus drinks and apple juice.)

A food scientist preparing a sample of orange juice for tests on nutritional content

COURSEWORK HINTS

- Page 58 introduces the food science coursework tasks – refer to this to remind yourself what is required.
- Practise the different food tests to make sure you know how to carry out each one effectively.
- Think carefully about which food tests you will need to carry out for this coursework task.
- Pages 156–7 contain the coursework checklist and mark scheme. Follow this advice.

COURSEWORK TASKS

The Food Standards Agency (FSA) is a government-funded organisation, whose role is to protect the public's health and consumer interests in relation to food.

Imagine you are a food scientist, and have been asked by the FSA to carry out a thorough investigation into the content of a new apple juice drink that has just been launched onto the market. The manufacturer has supplied the FSA with the nutritional content for this drink in the form of a food label. The drink carton also contains the claim that the drink contains 'at least 20% more vitamin C than orange juice'. Your job is to supply the FSA with an independent analysis of the drink, to ensure that what the manufacturer claims is correct.

Your coursework task is to carry out a series of appropriate food tests on the apple juice. You then need to write a report for the FSA, which compares the nutritional content of the drink (including either moisture content or suspended matter) against the information listed on the food label. Your report should also state whether or not the vitamin C content is as high as the packaging claims.

3.17 Investigating bacterial growth

LEARNING OBJECTIVE

1 How do you use aseptic technique, use streak plates and count colonies of bacteria to investigate bacterial growth?

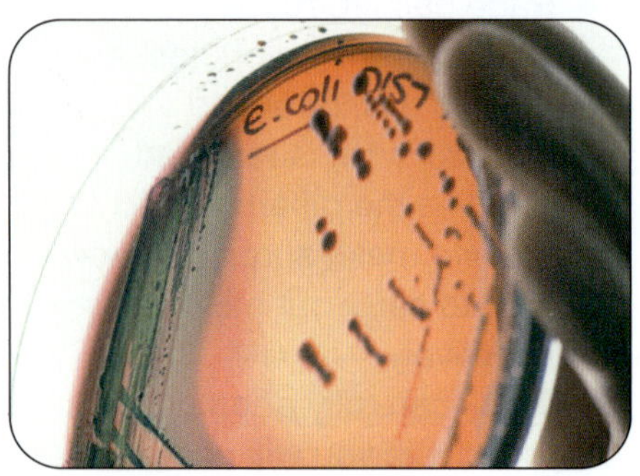

This is a streak plate of *E. coli* 0157. This *E. coli* strain was responsible for an outbreak of food poisoning in Scotland in 1996, when over a dozen people died and hundreds were taken ill.

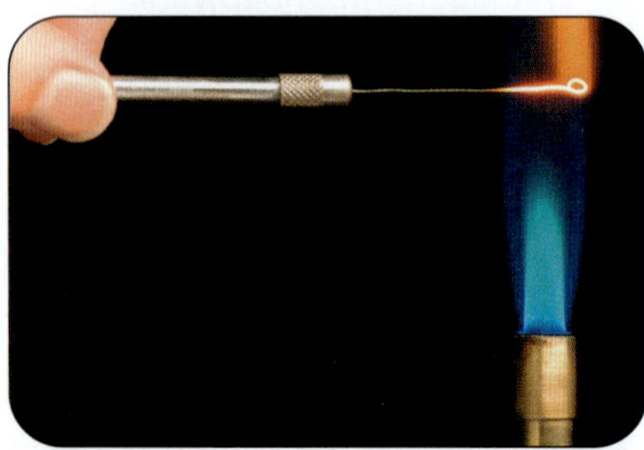
Sterilising a wire loop

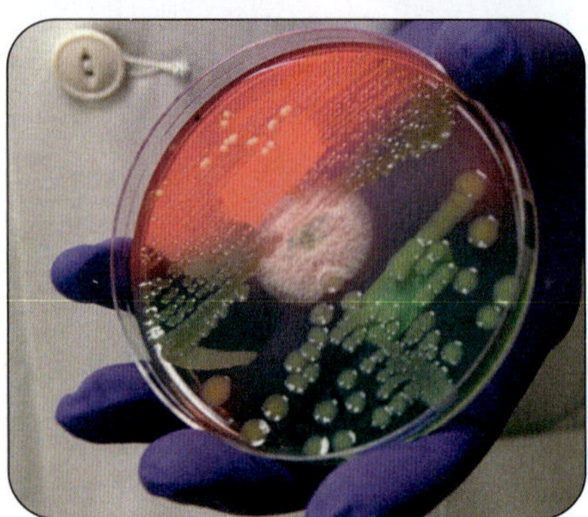
This laboratory technician is holding an agar plate containing several types of bacterium (yellow/green) and a fungus (white) cultured from an environmental swab

Food scientists and microbiologists are responsible for checking for the growth of bacteria in food and drink products. Sometimes scientists want to improve the growing conditions so that bacteria replicate as quickly as possible, e.g. in the production of yoghurt. More commonly, scientists want to prevent the growth of unwanted bacteria as they would spoil the food and could cause food poisoning. Techniques commonly used by microbiologists are:

Aseptic technique

Aseptic means 'without microorganisms'. Scientists use the aseptic technique to prevent unwanted microorganisms entering a sample they are studying, or passing from the sample to themselves, possibly causing disease.

Microorganisms are often transferred from one medium to another with a wire loop. Before the loop, or any other apparatus is used, it must be sterilised. To sterilise the loop, it should be heated until it glows red in a Bunsen burner flame. Before use it should then be allowed to cool. While cooling, the loop should be held close to the flame away from the bench to ensure it remains sterile.

a) What safety precautions should be taken when working with microorganisms?

Sampling the environment

Bacteria are commonly found throughout the environment. A sterile, cotton swab is used to sample bacteria on a surface such as a desktop, the floor, or a stair handrail. After rubbing the swab on the surface, it is then rubbed lightly on the surface of an agar plate. The lid is then put on and secured by four small pieces of tape ***(do not seal)***. The agar plate is then placed in an incubator at 25 °C for around 48 hours to allow any microorganisms present to grow, so that they can be identified.

b) Name some situations where scientists from the Food Standards Agency might want to check for the presence of bacteria.

Making a streak plate

This technique is used to isolate individual bacterial colonies, so that they can be identified. Bacteria can be recognised by the colony they form – they differ in characteristics like shape, colour, size and elevation.

1 Dip a sterilised wire loop into the sample of bacteria.
2 Make four or five streaks across one edge of an agar plate.
3 Flame and cool the loop.
4 Make a second series of streaks by crossing over the first set, picking up some of the cells and spreading them out across a new section of the plate.
5 Repeat steps 3 and 4 two more times making a third and fourth set of streaks. Fix lid with four short lengths of tape ***(Do not seal all the way round.)***
6 Incubate the plate upside down allowing the cells to form colonies. Do not open the plate. Dispose of plates in disinfectant or sterilise.

Serial dilutions to accurately count bacteria

It is extremely difficult to count the number of bacteria in a sample using a microscope, because there are so many bacteria to count. Diluting a bacterial sample spreads out the bacteria, so individual colonies of bacteria can be cultured when added to an agar plate. By counting the number of colonies that grow, the number of bacteria present in the original sample can be calculated:

Number of bacteria (per cm³) in original sample
= number of colonies × dilution of sample

c) A 1 cm³ bacterial sample was diluted to 1/100 000th of its original concentration to allow counting of colonies. If 46 colonies were counted, how many bacteria were present in the original sample?

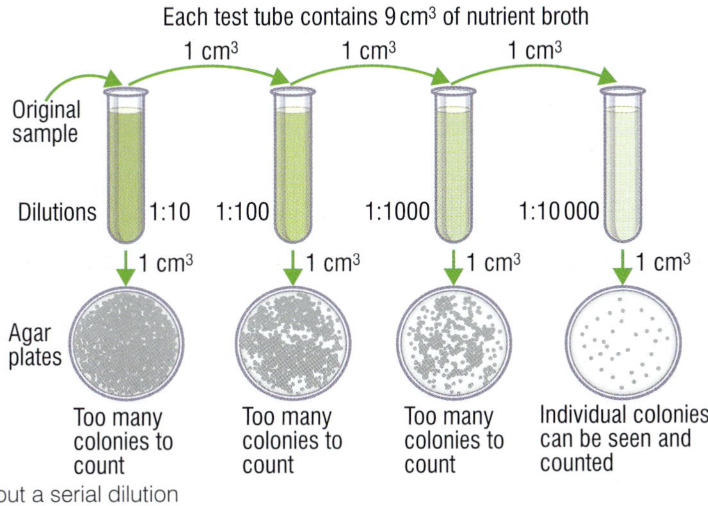

Carrying out a serial dilution

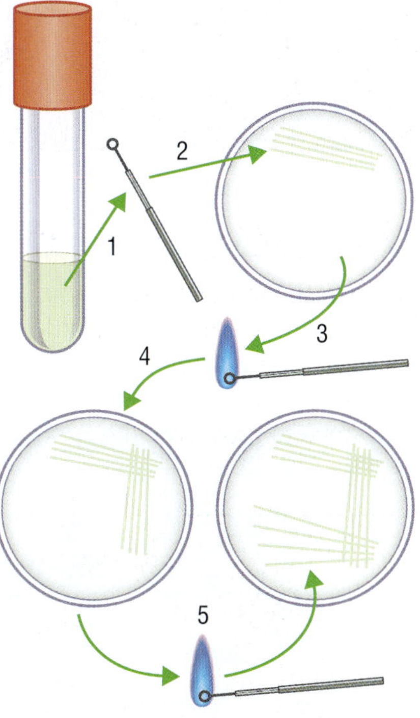

How to make a streak plate

COURSEWORK TASKS

Imagine that in the last 48 hours, 78 people have visited the local doctor complaining of sickness and/or diarrhoea, and a high temperature. Originally the doctor thought that a virus was causing these symptoms, but more detailed investigations showed that all these people had attended a wedding buffet two evenings ago in the village hall. An environmental health inspector was called to take samples of the food eaten; these were sent to a laboratory for analysis.

In the laboratory it was discovered that a strain of *E. coli* was responsible for causing an outbreak of food poisoning. Imagine you are a microbiologist who has been asked to determine what conditions of storage, preparation and cooking need to be adopted to prevent this microorganism from growing. You have been provided with a test tube containing nutrient broth and *E. coli*.

Your coursework task is to carry out a series of appropriate tests using *E. coli* to study the effect of different conditions on bacterial growth. These could include: temperature (***do not*** incubate at 37 °C), nutrient conditions, pH, level of moisture, and the presence of chemicals (e.g. detergents, disinfectants and antiseptics). You then need to write a report for the environmental health inspector, which includes advice to the catering company, explaining the practices that they should implement to ensure that no further outbreaks of food poisoning occur.

COURSEWORK HINTS

- Practise the different microbiological techniques to make sure you know how to carry out each one effectively.
- Think out carefully which techniques you will need to use to carry out this coursework task.
- Pages 156–7 list the coursework checklist and mark scheme. Follow this advice.

3.18 Investigating the effects of nutrients on plant growth

LEARNING OBJECTIVE

1 How do you investigate the minerals plants rely on for healthy growth?

Nitrate deficiency – older leaves are yellowed, growth is stunted

Phosphorus deficiency – younger leaves have a purple tinge and poor root growth

For healthy growth, plants need four important minerals:

- ***Nitrates*** (contain ***nitrogen***) – for healthy growth. Nitrates are involved in making DNA and amino acids. The amino acids join together to form proteins that are needed for cell growth.
- ***Phosphates*** (contain ***phosphorus***) – for healthy roots.
- ***Potassium*** – for healthy leaves and flowers.
- ***Magnesium*** – for making chlorophyll molecules.

In addition to these nutrients, 12 other elements are required in smaller amounts. Plants acquire these from the soil. When crops are harvested minerals are removed from the ground. These would normally be replaced when the plant dies, or when material such as leaves are shed.

Farmers need to be able to recognise mineral deficiency symptoms in plants so that they can add appropriate chemicals. These are normally in the form of a fertiliser. NPK is a common fertiliser, containing three of the essential minerals needed for healthy plant growth.

a) Summarise the symptoms of plant mineral deficiencies in a table.
b) How could farmers replace nutrients that are lacking in their soil?

Hydroponics

To ensure that a plant receives a measured amount of the minerals it needs, it can be grown in water with minerals dissolved in it. This system of growing plants is known as **hydroponics**. This is a technique often used by commercial growers because it enables plants to grow quickly.

c) Make a list of all the factors that can affect the growth of a plant. (Remember these factors will need to be kept constant if your investigation is to have meaningful results.)

Magnesium deficiency – leaves turn pale and then yellow

Potassium deficiency – yellow leaves with dead areas on them

Growing plants in a hydroponic system

Here is an outline of how plants could be grown in a simple hydroponic system:

1 Firstly, germinate several seeds in cotton wool. These should be kept damp and warm during the germination period.

2 When the seedlings reach about 5 cm in height, transfer them to conical flasks containing a nutrient solution.

3 Ensure that a wick is added between the solution and the cotton wool to keep the cotton wool moist, until the roots of the plant reach the nutrient solution.

4 Ensure that the plants have adequate light and are kept warm. Change the nutrient solution every 1–3 weeks.

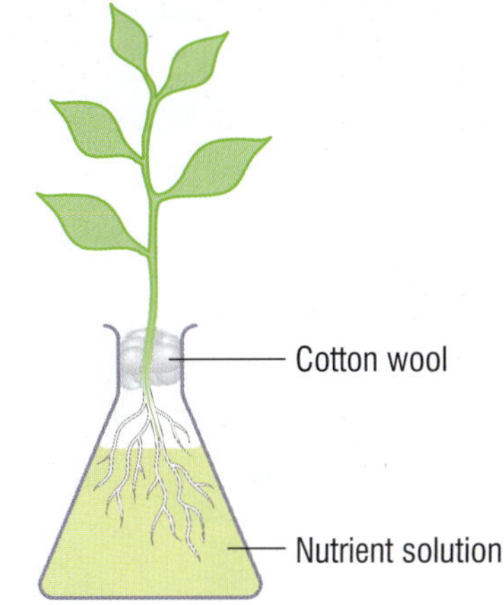

A simple example of a hydroponic culture

COURSEWORK TASKS

The well-known herbal specialist 'Bouquet Garni' have had a poor crop yield over the previous two years. The weather conditions at the farm have been similar to previous years, and no outbreaks of crop pests or disease have been reported. This has led the manager at the company to believe that the problem lies in the nutrient quality of the soil.

Imagine you are a crop scientist who has been hired by Bouquet Garni to help the company overcome this problem. They have asked you to investigate what is causing the poor growth of their crops, in order that they solve this problem before they begin planting for next season. They have provided you with the seeds they are going to sow, plus an example of a plant grown last season.

Your coursework task is to carry out an investigation into how a lack of specific nutrients can affect plant growth. Your investigation should include a control, with all nutrients supplied, plus plants grown in solutions deficient in specific nutrients.

You will need to decide when, and how, you are going to measure plant growth. In the photograph, scientists are measuring the height the plants have grown to. You could also measure:

- The length of the roots.
- The mass of the plant.
- The number of leaves.
- The average size of the leaves.

After completing your investigation you need to write a report to the manager of the company, explaining which nutrients the soil must contain to produce a bumper crop. Your report should also include details of the specific nutrient that was lacking at the farm, which led to the poor yield over the last two years.

The company has also asked you to include details of the effects different nutrient deficiencies have on plants. This will allow the manager to look out for these symptoms, to ensure a similar problem does not affect crop yields in future years.

These researchers are using a ruler and a laptop computer in an experiment into crop sizes. They are investigating the effects of herbicides and fertilisers.

COURSEWORK HINTS

- Practise growing a plant in a hydroponic system to make sure you know how to carry out this test effectively.
- See pages 160–1 for the coursework checklist and mark scheme. Follow this advice.

Forensic science

A SOCO (Scene of Crime Officer) photographing evidence

What you already know

Here is a quick reminder of previous work that you will find useful in this chapter:

- All substances are made of atoms. A substance that is made of only one sort of atom is called an element. Elements are shown in the periodic table.
- Atoms of each element are represented by a chemical symbol.
- Atoms have a small central nucleus around which there are electrons.
- When atoms react, their atoms join to form compounds. This involves giving, taking or sharing electrons and the atoms are held together by chemical bonds.
- Atoms and symbols are used to represent and explain what is happening to the substances in chemical reactions.
- The formula of a compound shows the number and type of atom that are joined together to make the compound.

From Key Stage 3:

- The pH scale is a measure of the acidity of a solution. Indicators help to classify solutions as acidic, neutral or alkaline.
- Metals and bases, including carbonates, react with acids.
- A chemical equation can summarise a reaction and shows what the products are.
- Solubility varies with temperature.
- The solubility of chemicals is different in different solvents.
- Light is refracted at the boundary between two different transparent materials.

RECAP QUESTIONS

1 a) What is the chemical symbol for the element sodium?
b) What group of the periodic table is sodium in?
c) What is the name and formula of the compound produced when sodium reacts with chlorine?

2 a) What is the charge on an electron?
b) What happens when a positively charged particle approaches a negatively charged particle?

3 a) What is the missing number below?
All acidic solutions have a pH value below ….
b) What are the products formed when hydrochloric acid and the alkali sodium hydroxide react?

4 a) What happens to the solubility of sugar in water as the temperature rises?
b) Name a solvent used in the home, other than water.

5 a) What is the difference between reflection and refraction?
b) What happens to light as is enters glass at an angle? Answer by drawing a diagram.

Making connections

Evidence

Evidence collected by crime scene investigators must link a suspect to the crime scene. Then the police present a successful case in court.

Impressions

Marks and impressions made by fingerprints, shoes, tyres and tools lead to the conviction of many criminals.

Chemical analysis

Chemical analysis of evidence includes the use of chromatography, flame tests and precipitation reactions.

DNA evidence

DNA evidence from blood, saliva, semen and urine is stored (like fingerprints) on a national database.

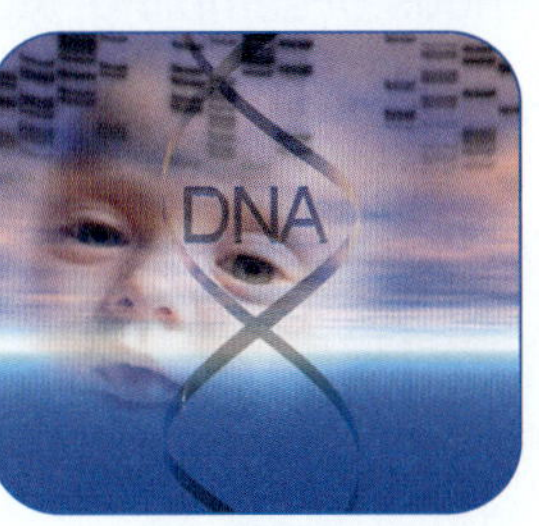

Trace evidence

'Every contact leaves a trace.' Trace evidence includes hair, fibre, paint, pollen, glass, dust and soil.

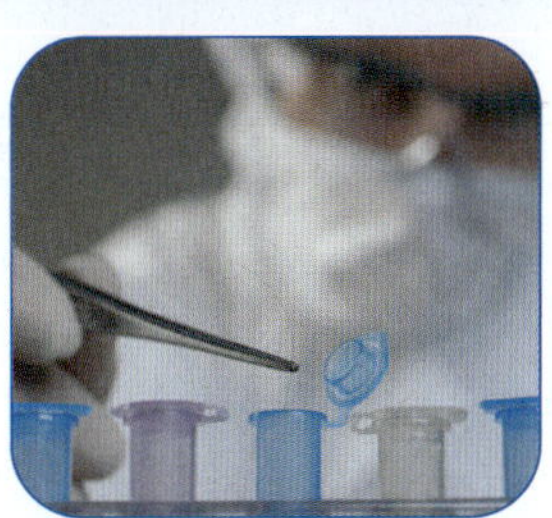

Methodical work

During the time between crime and court case, the forensic scientist is involved in patient, methodical work.

Microscopes

Microscopes are used to identify trace evidence and compare samples, such as matching a bullet to its gun.

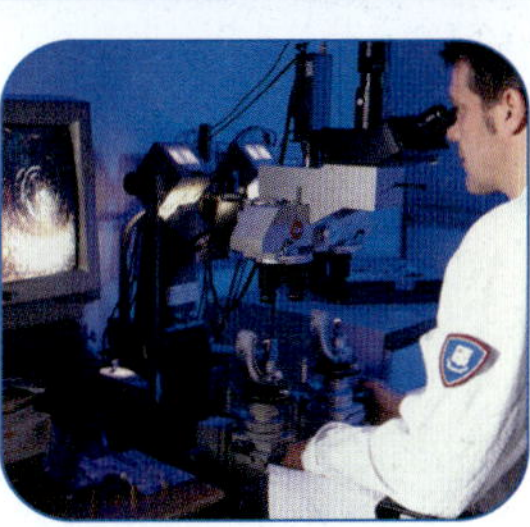

ACTIVITY

Debate the statement from a politician: "Levels of crime are dropping."
Discuss whether or not this is due to:

- Less crimes being reported.
- Statistics being doctored.
- More police on the beat.
- Better forensic techniques, leading to more successful prosecutions and deterring criminals.
- Targeted policing against drugs-related crime, reducing offences.
- A more socially responsible generation.
- A larger prison population, keeping career criminals off the street.

What do you think?

4.1 Introduction to forensic science

LEARNING OBJECTIVES

1 How do we use forensic science?
2 How do we avoid contaminating evidence at a crime scene?

In this chapter you will learn about some of the science and techniques used by forensic scientists.

Offences that forensic scientists investigate for the police include:

- Property crime – burglary, fraud, robbery, fires, theft of vehicles and theft from vehicles.
- Serious crime – murder, suspicious death and sexual offences.
- Drug-related crime – supply, possession, importation, production and cultivation.
- Road crime – drink- and drug-induced driving and accidents.
- Organised crime, including terrorism.
- Hi-tech crime – computer or electronic-related crime and fraud.

In their work, forensic scientists help the police to investigate and detect crimes, convict offenders and free the innocent. SOCOs are not referred to as forensic officers as they are employed by the police and are not police officers. Some police officers do have specialist forensic training.

Forensic work begins by carefully observing the crime scene and recording the materials found there. Only by collecting the samples properly will a court of law accept the evidence as valid or trustworthy.

Look at the photographs.

How do 'Scene of Crime Officers' (SOCOs) and 'Crime Scene Investigators' (CSIs) avoid contaminating the evidence by:

a) Restricting access.
b) Wearing protective clothing.
c) Using appropriate methods of sampling, storage and recording.

Notice the police tape to mark the limits of the crime scene. The SOCOs wear protective clothing, such as paper suits, overshoes, gloves and masks to prevent contaminating the crime scene.

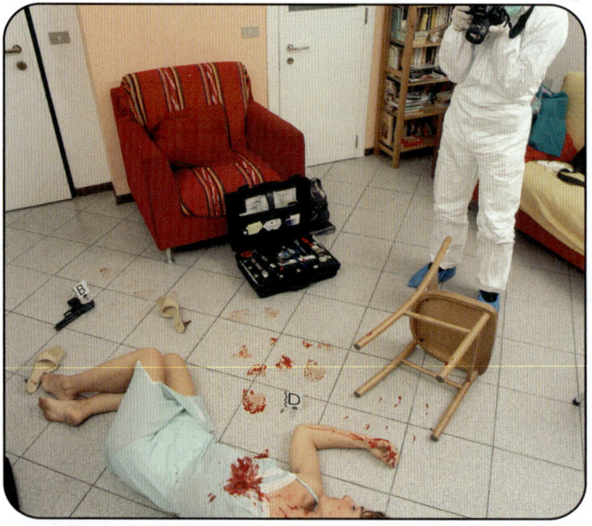

This is part of a police training exercise. A police officer acts as the victim. A forensic officer photographs the mock murder scene. This creates a permanent record for forensic investigations and court cases. Other photographs record the layout of the room and the location of evidence, such as blood traces, footprints, a gun and a spent cartridge shell.

Paper bags containing forensic evidence. See the yellow biohazard tape stuck on the packet. Paper packaging prevents condensation and contamination from bacterial growth.

Back in the laboratory, forensic scientists use a variety of scientific techniques to identify and match the samples obtained at the crime scene. Since we use the results of these tests as evidence in a court of law, accuracy and reliability are important.

Examples of evidence left at a crime scene include: fingerprints, hair, fibres from clothes, footprints, glass fragments, chemical compounds, blood and DNA evidence.

The 'National DNA Database' (NDNAD)

When a suspect is arrested, the police have the power to take a mouth swab. Forensic scientists analyse cells from this swab and develop a '***DNA profile***'. They put this profile on the 'National DNA Database'. Next they check this profile against profiles obtained from scenes of crime. Any matches between the suspect and a crime scene could lead to a successful police prosecution.

In this way, suspects' DNA profiles have been matched to those from a whole range of crimes – from murder, manslaughter and rape, to burglary and car crime.

DID YOU KNOW?

DNA stands for 'deoxyribonucleic acid'.

DNA is the chemical that carries genetic information in the form of a code. It determines your physical traits. Half of your DNA is inherited from your father and the other half from your mother. Except for identical twins, each person's DNA is unique. We can extract DNA from blood, semen, saliva or hair samples.

SUMMARY QUESTIONS

1 What do: a) 'SOCO' b) 'CSI' and c) 'NDNAD' stand for?

2 How is completing a jigsaw puzzle like solving a crime?

3 Look at the photograph. When collecting the cigarette butt, how could an inappropriate technique lead to uncertainty about the validity and reliability of the evidence?

4 We use forensic science for other purposes besides helping the police investigate crimes. Give an example of the use of forensic methods when:

a) Studying archaeological specimens.
b) Investigating the cause of an industrial accident.
c) Showing whether or not people are related.

A SOCO, using a pair of tweezers, collects a cigarette butt found at a crime scene. The gloves and protective clothing help to prevent contamination of the evidence. He will analyse the cigarette butt for traces of skin cells and saliva. DNA profiling may then link a suspect to the crime scene.

KEY POINTS

SOCOs and CSIs must protect a crime scene and avoid contaminating evidence by:

1 Restricting access with tape to mark out the boundary of the crime scene.
2 Wearing protective clothing, such as paper suits and overshoes, gloves and masks.
3 Using suitable methods of collecting and storing samples, and recording information, such as labelling evidence bags and taking photographs.

4.2 Fingerprints

LEARNING OBJECTIVES

1. What are the three distinctive types of fingerprint pattern?
2. How do we reveal fingerprints on different surfaces?

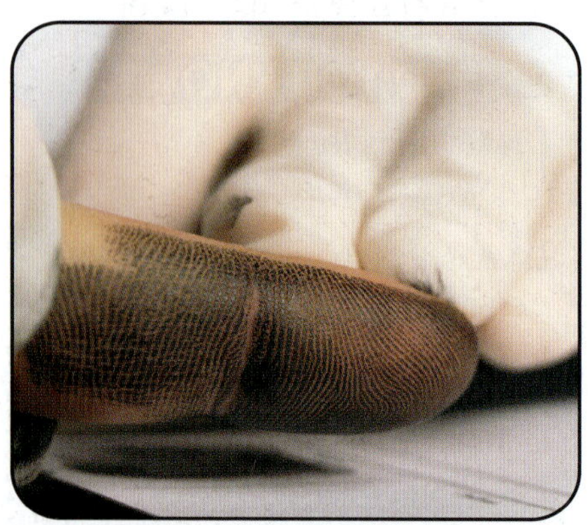

Fingerprinting: rolling the suspect's finger over the paper

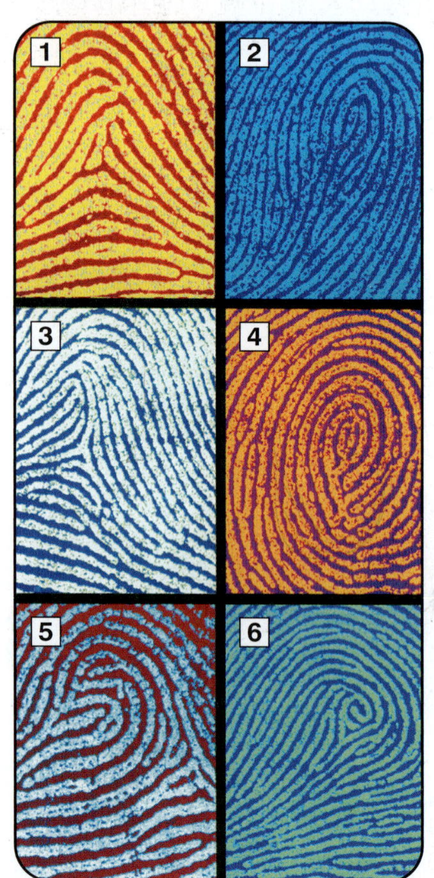

No two fingerprints are the same. A fingerprint is unique. If it's there, so was the person to whom it belongs.

The National Automated Fingerprint Identification System (NAFIS) allows rapid matches to be made between prints found at the crime scene and those on the database.

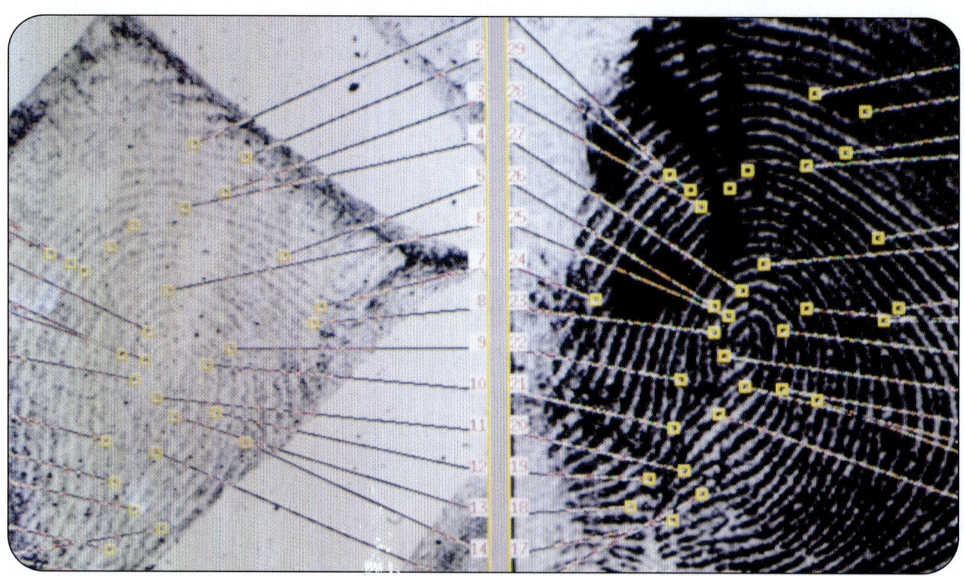

A fingerprint is a deposit of oils and chemicals left after sweat has evaporated. The pattern provides a mirror image of the ***ridges*** and troughs of the finger.

Criminals leave fingerprints at 30% of crime scenes, particularly their point of entry. Other places to check are open drawers and objects that have been moved. Once you find a fingerprint, it is essential to photograph it.

a) In a burglary, where is the obvious place to start looking?
b) Look at the three main types of ridge: arches, whorls and loops. Identify the types of fingerprint (1–6) in the photograph (an arch, a whorl or a loop).

Arch, 5% of patterns

Whorl, 25% of patterns

Loop, 70% of patterns

The common ways to recover a fingerprint are:

- ***Powder dusting*** – Carbon black is used for light surfaces, aluminium for dark surfaces and magnetic powder for porous surfaces and plastics.

PRACTICAL

1. Gently brush the fine powder over the fingerprint, causing the dust to stick to the oil in the ridged pattern.
2. Press low-tack lifting tape to the print. Carefully lift the fingerprint from the surface.
3. Place on a fingerprint card and label.

- ***Quaser illumination*** – Shining a bright white light can reveal a print. Exposing the fingerprint to violet light causes the oils in the ridges to glow (fluoresce). Spraying dyes that react with the natural oils of the fingerprint enhances its appearance. For example, ninhydrin-treated prints appear purple, while DFO (1,8-diazafluoren-9-one) will fluoresce under violet light.
- ***Superglue fuming*** – Exposing an object to fumes of heated Superglue enhances a fingerprint. The Superglue bonds with the grease in the ridges to reveal a white fingerprint. Dyes improve the contrast and make the print even more visible.

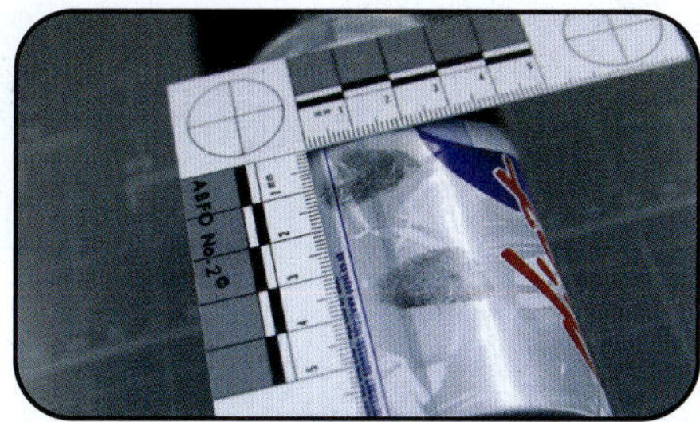

Photographing a drink's can as evidence. Scales, stuck to the can, enable actual size photographs to be produced.

DID YOU KNOW?

In 1998, a fingerprint at the bottom of a swimming pool solved a murder.

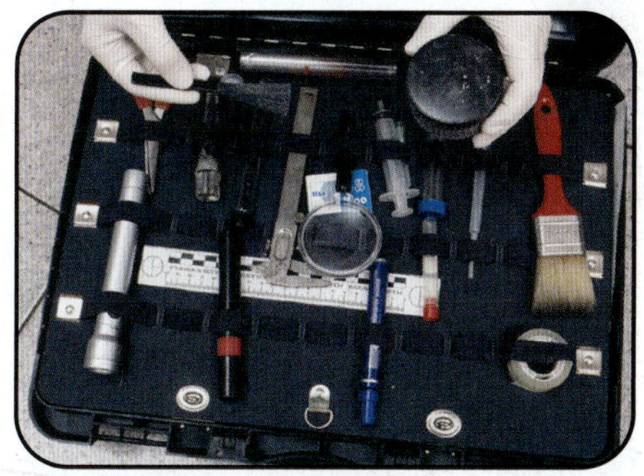

Crime scene investigation kit

SUMMARY QUESTIONS

1. What is the shape of the ridges on your left forefinger: arch, whorl or loop?
2. Look at the white Superglue prints on this perfume bottle. Why could dye make the fingerprints clearer?

Superglue fuming

Magnetic powder

3. a) When dusting with magnetic powder, the dust clings to a magnet. There is no brush and no bristles.
   Why could this help reveal clearer fingerprints?
   b) How is a magnetic 'wand' useful for collecting up unused powder?
4. Fingerprints found at the crime scene enable the police to link a suspect to a crime. What else do the police need to be able to charge the suspect?

KEY POINTS

1. There are three main types of fingerprint: loops, arches and whorls. (LAW for short!)
2. The common ways of recovering fingerprints are:
   - Powder dusting.
   - Quaser illumination: Violet light causes oils in the ridges to fluoresce.
   - Superglue fuming: Superglue bonds with the oils in the ridges.

4.3 Glass – the invisible evidence!

LEARNING OBJECTIVES

1 How can trace evidence, like minute fragments of glass, be matched to a suspect?
2 How does a glass surface refract light?
3 What is the refractive index of a glass fragment and how do we measure it?

Using forceps to collect a 'control glass sample' from a broken window. Will this sample match with fragments on the clothes or in the hair of a suspect?

Edmond Locard was known as the Sherlock Holmes of France. According to Edmond Locard, no criminal is invisible. "Every contact leaves a trace." The guilty always take identifying evidence away with them. They always leave traces of evidence at a crime scene – hairs, paint flecks, pollen, soil, fibres and tiny flakes of glass.

We find tiny slivers of glass on suspects:

- Vandalising or breaking and entering through windows.
- Involved in assault with bottles.
- From their vehicle headlights in hit-and-run road traffic incidents.

It is sometimes possible to match larger glass fragments together like a jigsaw. But how can we link a suspect to the crime scene with just a few slivers of glass? **Refractive index** (or light-bendability) can help us out. Different types of glass have different densities and so, bend light by different degrees. We can find the refractive index of glass by measuring the angle of incidence (i) and the angle of refraction (r).

PRACTICAL

Which refracts more, glass or Perspex?

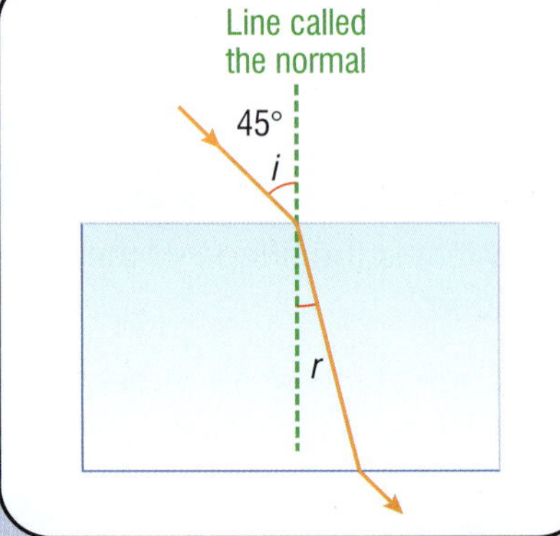

Refraction

1 Draw round your glass block.
2 Use a protractor to mark a line at 45° as shown.
3 Direct a ray from a ray box along this line, then mark the ray that emerges from the glass.
4 Remove the glass block and complete the diagram as shown.
5 Measure the angle of refraction r.
6 Calculate the refractive index using the formula:

$$\textbf{Refractive index} = \frac{\sin i}{\sin r}$$

7 Repeat using the Perspex block.

- Which material has the greater light-bendability?

a) What equation do we use to calculate refractive index?

We need a different method for measuring the refractive index of glass fragments.

Strangely, forensic scientists make glass fragments invisible to detect their refractive index!

Look at the photograph of the liquid droplet. Suppose the glass of the pipette and the liquid have the same refractive index (light-bendability). Then the light would not change direction at the glass–liquid surface. So you could not tell where the glass started and the liquid ended. As if by magic the surface is invisible!

A liquid droplet hanging on a pipette

Oil immersion method

To measure the refractive index of a tiny bit of glass, a forensic scientist immerses the fragment in an oil, under a microscope. The scientist then slowly heats and cools the oil. This changes the oil's refractive index. At just the right temperature the glass disappears – when the refractive index of glass and oil match. A computer converts the temperature into a refractive index value.

b) The glass at a crime scene and the glass on a suspect have the same refractive index. Does this prove the suspect's guilt? Explain your answer.

PRACTICAL

Invisible glass

1 Use a glass pipette. Fill three test tubes with glycerol.

2 Warm two in hot water, taking one out earlier. Your glycerol is now at three different temperatures.

3 Immerse your pipette into each. Does the glass of the pipette all but disappear?

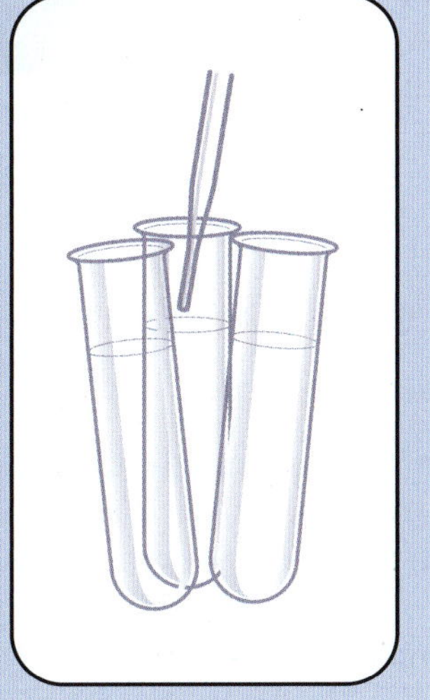

PRACTICAL

Refraction with a sugar cube

Suspend a sugar cube in water. Illuminate with light. As the sugar dissolves the denser sugar solution falls to the bottom. The solution has a different refractive index to the water, as its light-bendability shows.

SUMMARY QUESTIONS

1 Which way does the light refract (A–E)?

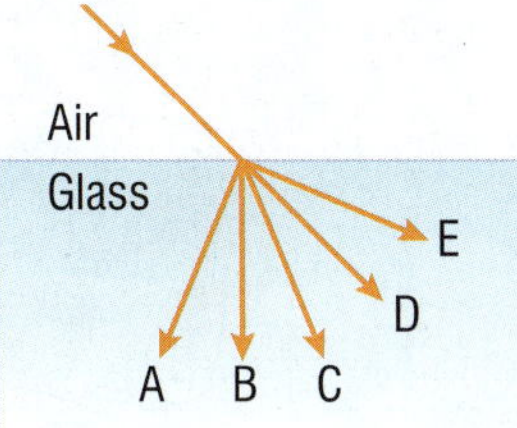

2 How does the diagram explain Locard's principle?

"Every contact leaves a trace."

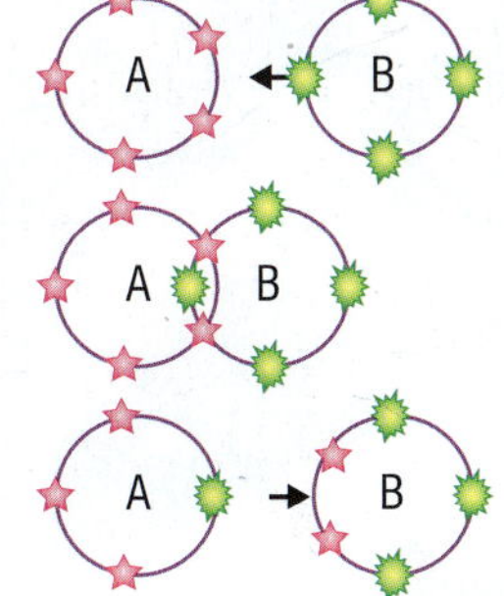

3 In an experiment the angle of incidence is 60° and the angle of refraction is 30°. What is the refractive index of the material?

4 A suspect has just washed his clothes. You comb 11 glass fragments from his hair. All but one match the refractive index of the control sample found at the crime scene. What can you conclude?

KEY POINTS

1 Light refracts (bends) towards the normal when it enters glass.

2 We can find the refractive index by measuring the angle of incidence and the angle of refraction.

3 Oil immersion method: a glass fragment disappears when put in oil with the same refractive index.

4.4 Microscopic evidence

LEARNING OBJECTIVES

1. Why are instruments in forensic laboratories likely to give more accurate evidence than those in school or college laboratories?
2. What are the distinctive features of pollen grains and layers of paint?

Many types of evidence are far too small to see with the naked eye. Although forensic scientists use standard light microscopes, such as we have in schools and colleges, they also use more expensive and powerful microscopes. These give better ***resolution*** and ***contrast***.

Resolution

The better the resolution the sharper the image.

With a ***scanning electron microscope*** (SEM) we can see objects as small as 10nm (10 nanometres). 1 nm is one millionth of one millimetre, as small as many molecules. SEMs give a 20x better magnification than the best light microscopes.

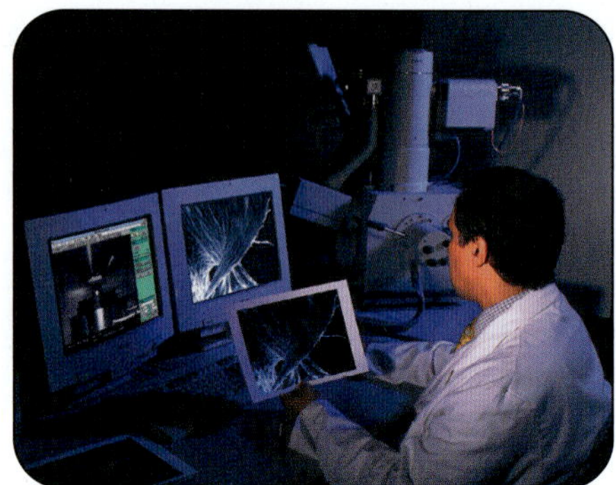

A technician holds the scanning electron microscope (SEM) image of a moss, found on the clothes of a suspect

Contrast

The better the contrast the clearer the image.

We can improve contrast by using:

- A ***stain*** – A coloured dye can make hidden detail clear.
- A ***comparison microscope*** – This is actually two microscopes side-by-side. We then compare the two images, on a computer screen, to see if they match.

Used bullet cartridges in two microscope holders for comparison. Unique markings are made when the firing pin hits the back of a shell. We want to compare marks on a cartridge recovered from a crime scene, to marks on a shell test-fired from the suspect's gun.

a) What does it prove if these markings match?

- A ***polarising microscope*** – You may know that wearing Polaroid glasses reduces glare. Placing paint pigments and fibres between crossed-Polaroids can brighten the object while darkening the background.

Collecting paint evidence. Layers show a car has recently been re-sprayed. The paint is compared with known paint samples.

b) Tell a story based on the three photographs.

The resolution and contrast of these microscopes provide more accurate and reliable evidence than the microscopes used in schools and colleges.

Trace evidence

Trace evidence often links a suspect to the scene of a crime. Remember, "every contact leaves a trace". Paint flecks and glass fragments, dust and soil, fibres and hairs, seeds and pollen, can all transfer to and from the clothing of an offender. For example, paint flecks transfer when burglars break and enter, vandals scribble graffiti, and cars crash in road traffic accidents.

Pollen

Pollen grain from plants at the crime scene can transfer to clothing. Washing may not remove them from the seams of the garment. The main features that distinguish one type of pollen from another are size, shape and surface texture. The outer walls have pores and furrows. They can be meshed, granular, spined, or smooth.

Dust

Dust contains shed skin cells, dust mites, hairs and fabric threads, soil and earth, pollen grains, fungal spores and particles of food. Three or more hairs fall unnoticed from our heads every hour.

Scene of crime officers (SOCOs) use tape or suction tubes to lift samples of dust and dirt from surfaces such as carpets and car seats. This evidence is unique to the crime scene. The tell-tale fibres that garments shed are also distinctive.

Dust from a carpet. A combination of different multicoloured fibres, from a garment or a carpet, is conclusive evidence to match a suspect to a crime.

SUMMARY QUESTIONS

1 Which clothing fibre, a, b or c, is the best match to the control sample?

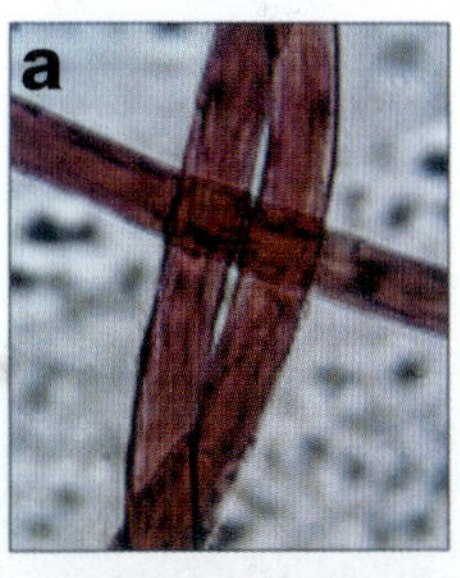

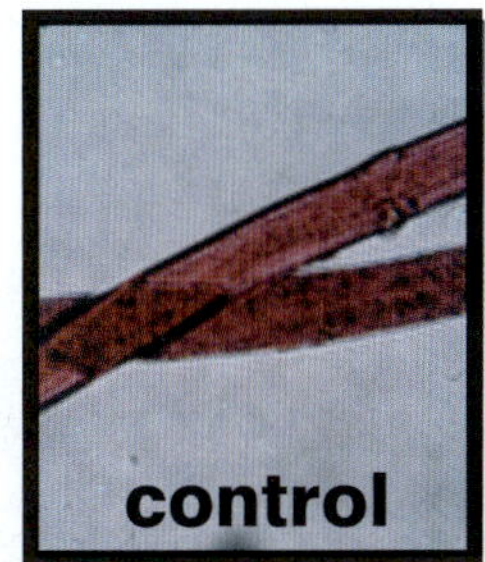

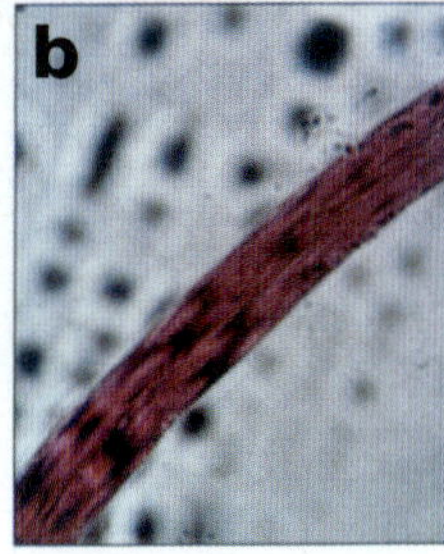

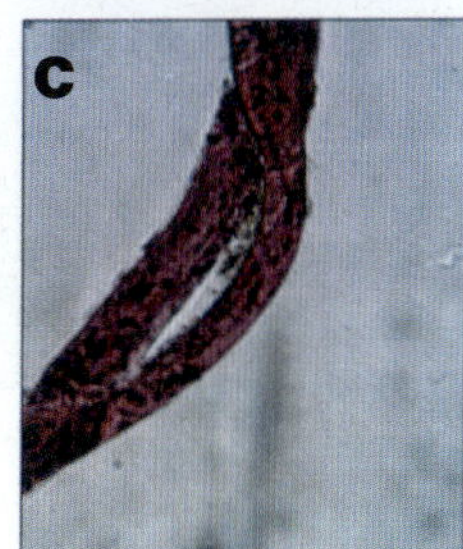

2 Suppose you collect paint flakes at a crime scene and find some on a suspect. They do not seem to fit together, yet layers observed under a microscope are identical. What does this prove?

3 Why must all evidence be bagged, sealed and labelled?

KEY POINTS

1 Forensic scientists use comparison, polarising and electron microscopes to improve contrast and resolution. This provides more accurate and reliable evidence than simple light microscopes.

2 The pores and furrows of pollen grains and layers of paint have distinctive features.

3 Trace evidence can link a suspect to a crime scene.

4.5 Impressions

LEARNING OBJECTIVES

1 Which distinctive marks can we compare?
2 How do we take samples of marks or impressions?

DID YOU KNOW?

Distinctive bite marks in an apple convicted an IRA hit man of murder.

Offenders often leave marks and impressions at the crime scene, such as footprints, tyre tracks and tool marks. You can ***photograph*** these to provide a permanent record of the evidence.

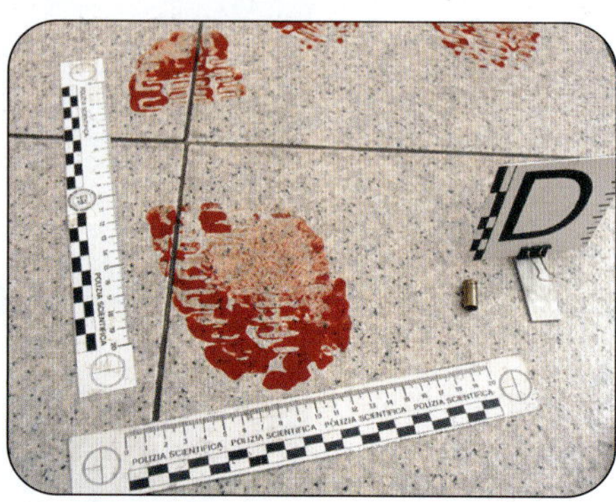
Photograph of bloody footprints. Rulers provide a scale.

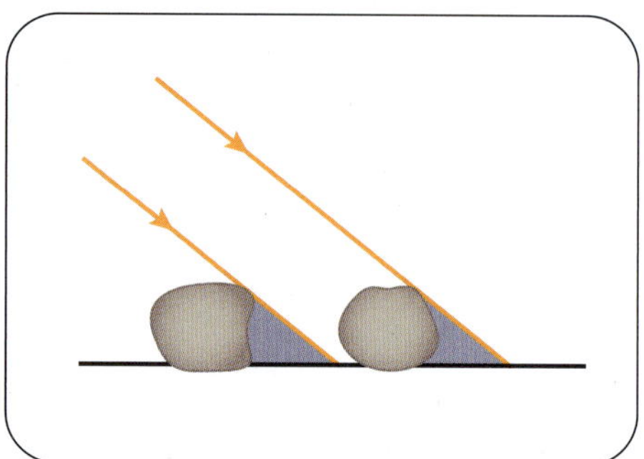
Dust particles. You hold a bright light at an oblique angle when searching for evidence. This also helps when photographing footprints.

As time passes, impressions like footprints and tyre tracks may be destroyed. Rain, wind, or animals and people walking over and disturbing the impression can all affect it in some way.

a) How could footwear impressions be ruined?

The following practicals below show two methods of recovering footprints at the crime scene.

Impressions of a shoe and a car tyre

PRACTICAL

Making a plaster cast of the footprint

1 Make a cardboard support round the footprint.
2 Mix plaster of Paris with water.
3 Pour the mixture into the mould.
4 Allow time for the cast to set.

(Just as dentists take impressions of teeth, forensic scientists take impressions of footprints. In fact they prefer dental stone to plaster of Paris as it shows detail more clearly.)

Footprint recovered by making a plaster cast

b) Compare casting for footprints and tyre tracks.

PRACTICAL

Electrostatically lifting the surface of dust from the footprint

The high voltage induces a charge on the particles of dust. The dust attracts up to the foil above. The black plastic surface is a good contrast in colour to the lighter coloured dust.

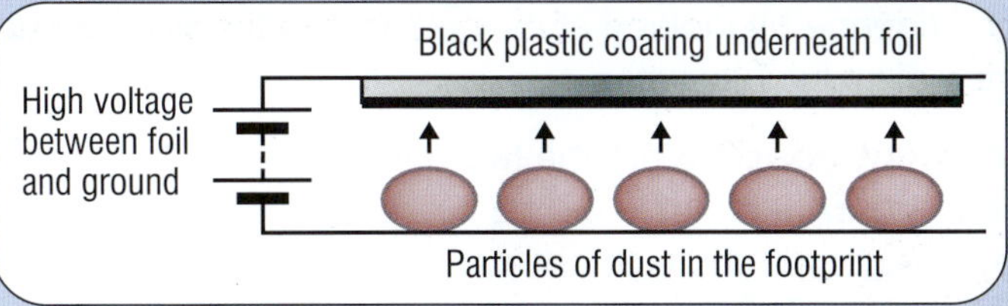

Electrostatically lifting the two-dimensional surface of dust from the footprint

c) After electrostatic lifting, why must we take a photograph?

The whole idea of impression comparison is that you retrieve both the impression ***and*** the object. Their size and pattern, their degree of wear and damage from cuts and scratches, make shoes and tyres unique. These distinct details can reveal a match, linking a suspect to the crime scene. Once you recover a suspect's footwear, you can make an impression by ***inking*** the sole and pressing the shoe onto a sheet of paper.

Forced entry or physical violence, using a tool, produces surface indentations. In either case you photograph the tool mark and make a mould of the impression using either dental stone, putty or silicone rubber.

d) Why is Plasticine not such a good material for making a mould?

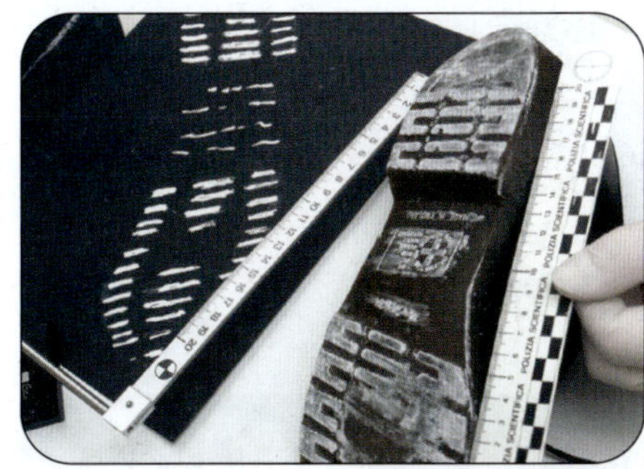

Measuring footprints. A comparison with footprints from the crime scene may link the suspect to that crime scene. Footprints at a crime scene can also indicate the number of people present, their weight and direction of movement.

Labelling evidence

It is important to label evidence with:

- What the evidence is.
- Its location at the crime scene.
- The date and the signature of the person collecting the evidence.

You must also keep an audit or record of what happens to evidence, from its discovery at the crime scene, to its use in a court of law. This dispels claims of ***evidence tampering***.

KEY POINTS

1 Footprints, tyre tracks and tool marks are examples of impressions.
2 To record impressions at the crime scene make a plaster cast and take photographs.
3 When you recover footwear from a suspect, make an impression by inking and taking photographs.

SUMMARY QUESTIONS

1 What tools might a criminal use to force entry to a home?

2 State four ways to obtain impressions of footprints.

3 Look at the marks on the bullets. One was found at the crime scene. Scientists fired the other from a suspect's gun. Comparing the two bullets, what do you conclude?

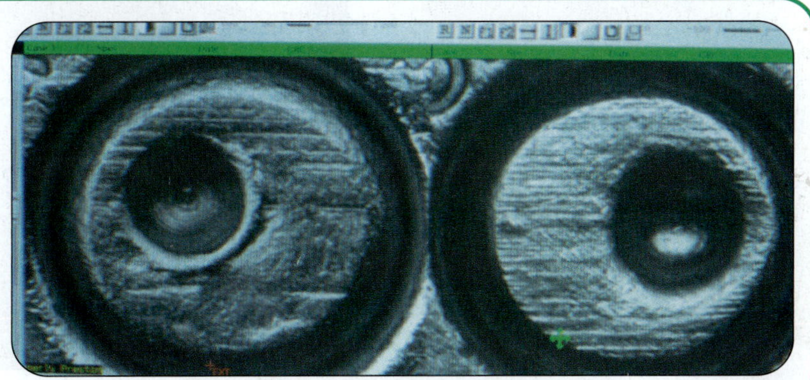

4.6 Distinguishing different substances

LEARNING OBJECTIVES

1. How do forensic scientists identify substances from a crime scene?
2. What are the properties of ionic compounds?
3. Why do many covalent compounds have low melting and boiling points?
4. What is the formula of certain ionic and covalent compounds?

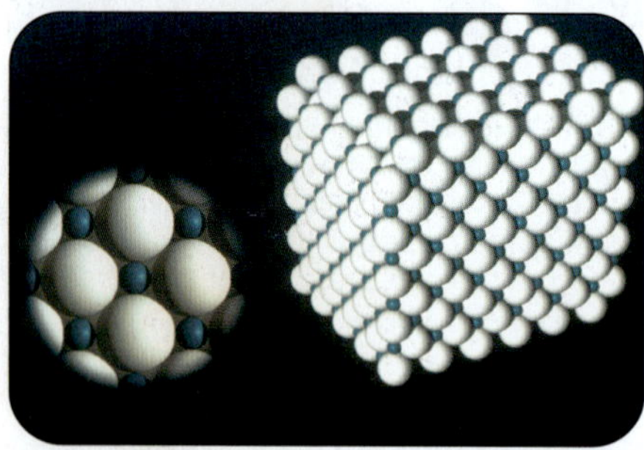

Cubic crystals of sodium chloride

Forensic scientists carry out tests on substances to use as evidence. They can identify a substance for evidence by looking at its melting point, boiling point and how it behaves when dissolved in water.

Ionic bonding and **covalent bonding** are the two different ways that atoms can join chemically. A compound with ionic bonds has a high melting point and usually dissolves in water. A compound with covalent bonds usually has a low melting point and is often insoluble in water.

Ionic bonding

An **ion** is an atom that has gained or lost an electron. **Electrons** are the tiny negative particles orbiting in an **atom**. An ionic bond occurs when:

- A metal atom loses an electron(s) and becomes a positive ion.
- A non-metal atom gains an electron(s), becoming a negative ion.
- Many millions of these ions attract and bond together in giant structures.

The small spheres in the photo represent positively charged sodium ions. The large spheres are the negatively charged chloride ions. ***Electrostatic attraction*** between neighbouring ions, operating in all directions, holds the giant lattice together. This makes the structure strong. To break the strong bonds between positive and negative ions requires a lot of energy. This explains why ionic compounds, like sodium chloride, have high melting points.

In ionic compounds:

- Group 1 metals, like sodium (Na), form 1+ ions.
- Group 2 metals, like magnesium (Mg), form 2+ ions.
- Group 6 non-metals, like sulfur (S), form 2– ions.
- Group 7 non-metals, like chlorine (Cl), form 1– ions.

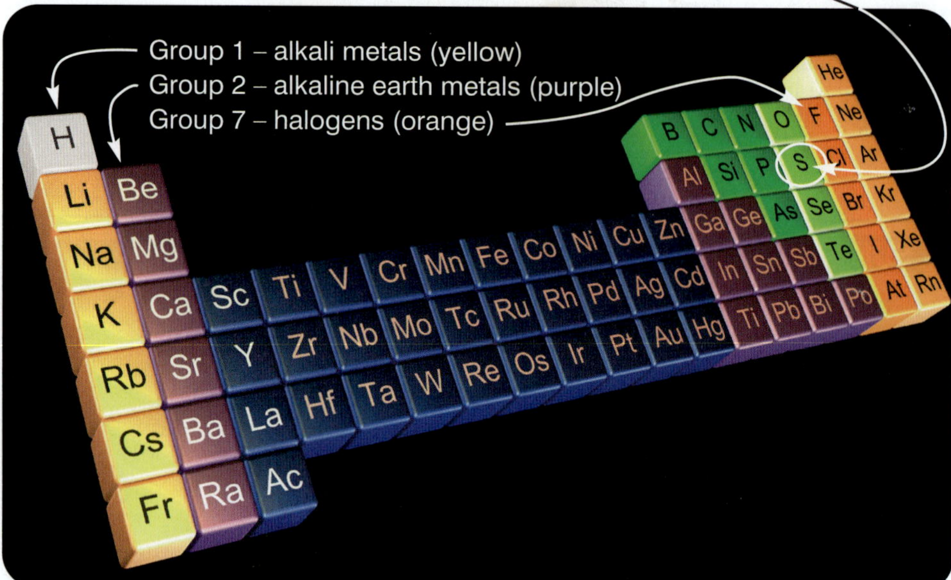

The periodic table. Some groups (1, 2 and 7) are colour-coded.

Writing formulas for simple ionic compounds

Ionic compounds have no overall charge.

So to work out a formula, ***match the charges*** so that the positive and negative charges cancel each other out. However, we write the formula without the charges.

e.g. sodium chloride NaCl (1+, 1–)
magnesium oxide MgO (2+, 2–)
sodium sulfate Na_2SO_4 (2+, 2–)
(We need 2 Na^+ to cancel the charge on SO_4^{2-}.)
copper(II) nitrate $Cu(NO_3)_2$ (2+, 2–)
(The (II) shows Cu(II) has a 2+ charge.)

a) What is the formula of copper(II) bromide?

Metal ions		Non-metals ions	
After losing 1 electron:		*After gaining 1 electron:*	
sodium	Na^+	chloride	Cl^-
potassium	K^+	bromide	Br^-
silver	Ag^+	nitrate	NO_3^-
		hydroxide	OH^-
After losing 2 electrons:		*After gaining 2 electrons:*	
magnesium	Mg^{2+}	oxide	O^{2-}
calcium	Ca^{2+}	sulfide	S^{2-}
iron(II)	Fe^{2+}	sulfate	SO_4^{2-}
copper(II)	Cu^{2+}	carbonate	CO_3^{2-}
zinc	Zn^{2+}		
lead	Pb^{2+}		

Covalent bonding

When non-metal atoms join together they form covalent bonds. They do this by sharing electrons. This creates strong bonds between the atoms ***within*** each **molecule**, e.g. oxygen O_2, water H_2O, ethanol C_2H_5OH and glucose $C_6H_{12}O_6$.

Yet the forces of attraction ***between*** the molecules are weak. Covalent compounds have low melting points because we only need a little energy to separate the molecules from each other.

b) Why is carbon dioxide a gas at room temperature?

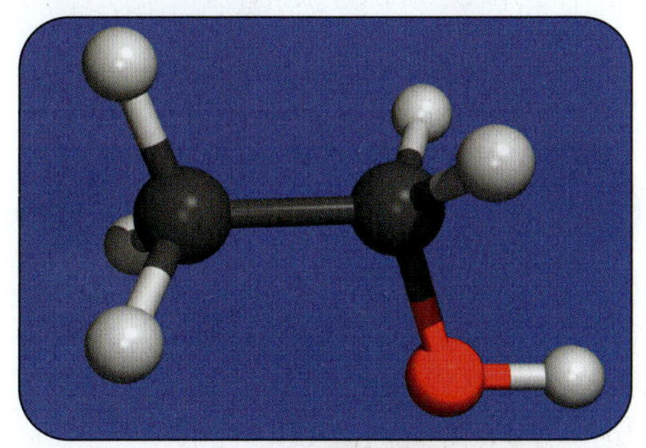

An ethanol molecule (C_2H_5OH or CH_3CH_2OH) showing two carbon atoms (black), six hydrogen atoms (grey) and one oxygen (red). The 'sticks' show the covalent bonds.

Substances obtained from living materials are **organic** compounds with covalent bonding. We produce ethanol by fermenting sugar with yeast for alcoholic drinks. We also use ethanol for fuel and as a solvent in industry and in laboratories.

Tests:

- You can see the test for ethanol on pages 104–5.
- Glucose ($C_6H_{12}O_6$) is another covalently bonded molecule that you need to know the test for. (See page 62.)

SUMMARY QUESTIONS

1 What do positive ions and negative ions do to each other?

2 What sort of bonding is there in a giant lattice, held together by strong electrostatic forces of attraction?

3 What sort of bonding is there in an organic compound?

4 Which is more likely to dissolve in water, an ionic compound or a covalent compound?

5 Explain why marble (calcium carbonate) has a high melting point, but water has a low melting point.

6 What is the formula of each of these molecules?
a) chlorine b) carbon dioxide c) calcium oxide d) zinc chloride
e) lead sulfate f) sodium carbonate g) potassium sulfate h) iron(II) nitrate

KEY POINTS

1 The properties of substances can be used to help identify the substances.
2 Ionic compounds have high melting points, while most covalent compounds have low melting points and boiling points.
3 Ionic compounds have ionic bonds between positive metal ions and negative non-metal ions, as in sodium chloride NaCl. To break the bonds needs a lot of energy.
4 Organic compounds have covalent bonds, like glucose $C_6H_{12}O_6$. To separate the molecules only needs a little energy.

4.7 Testing for ions

LEARNING OBJECTIVES

1. How do we analyse chemical evidence from a crime scene?
2. What are precipitation reactions and how are they used?

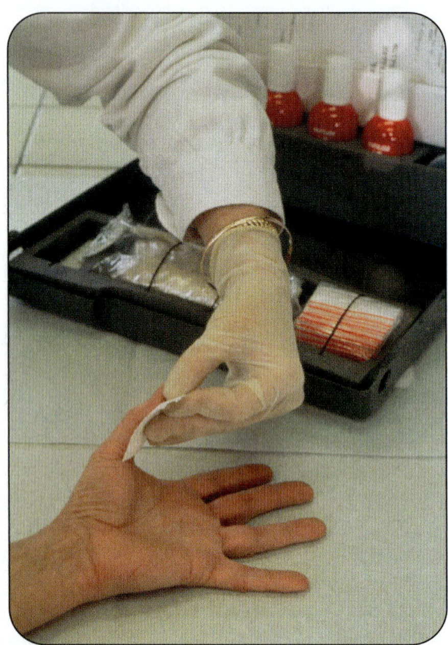

Wiping the hand of a suspect. The swab will be analysed for the presence of chemicals.

A technician using a mortar and pestle to grind a soil sample. She will use standard tests to analyse for metal ion contamination.

Forensic scientists often have to analyse chemical evidence found at the crime scene and on suspects' clothes and hands. They always aim for the most appropriate, yet simplest method to carry out a task. It is not necessary to waste time or money.

PRACTICAL

Analysing ionic compounds

The identity of a chemical may provide us with vital evidence. You have probably seen chemical reactions in which ions seem to 'swap partners':

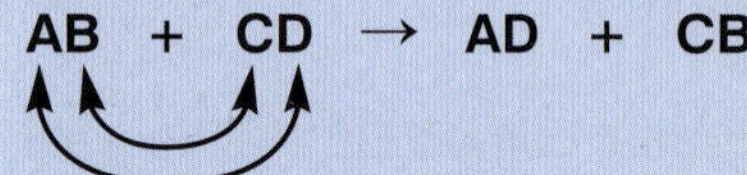

Here is an example:

lead nitrate + potassium iodide → lead iodide + potassium nitrate

$$Pb(NO_3)_2 + 2KI \rightarrow PbI_2 + 2KNO_3$$

Notice that solutions of lead nitrate and potassium iodide are both colourless. Now, wearing eye protection, mix them together in a test tube, one drop at a time.

- What happens?

The lead iodide forms as a **precipitate**. The lead iodide does not dissolve. We call this a **precipitation reaction**.

Safety

Chemicals you are using are toxic, corrosive and harmful / irritant, so take care and wear eye protection.

Precipitation reactions

In a precipitation reaction, a suspension or insoluble solid forms when two solutions react together. From the colour of the solid produced, you can tell the name of one of the reactants.

You use precipitation reactions to analyse the contents of solutions. The solution may come from a polluted river and you need to find out who is responsible for the pollution. You need two tests, one to test for positive metal ions and another to test for negative non-metal ions. Remember to dispose of waste correctly.

PRACTICAL

The sodium hydroxide (NaOH) test for positive metal ions

1 Wearing eye protection and taking care, put 2–3 cm^3 of the following sample solution into a test tube.

2 Add a few drops of 1 mol/dm^3 sodium hydroxide solution.

Metal ion in solution	Observation after adding NaOH solution
Aluminium (Al^{3+})	White cloudy jelly-like precipitate, which dissolves if we add more NaOH.
Lead (Pb^{2+})	White precipitate, which dissolves if we add more NaOH.
Calcium (Ca^{2+}), Magnesium (Mg^{2+})	White precipitate, which doesn't re-dissolve.
Copper(II) (Cu^{2+})	Blue-green, jelly-like precipitate.
Iron(II) (Fe^{2+})	Green-grey, jelly-like precipitate.
Iron(III) (Fe^{3+})	Red-brown, jelly-like precipitate.

In each of these reactions the precipitate is the metal hydroxide, e.g.

lead nitrate + sodium hydroxide → <u>lead hydroxide</u> + sodium nitrate

a) What colour is copper hydroxide?

Test: See page 84 to see how to distinguish Na^+ from K^+, and Ca^{2+} from Mg^{2+} ions.

PRACTICAL

Tests for negative non-metal ions

The following table lists the tests for negative metal ions. Carry out each of these tests and note the observation.

Non-metal ion in solution	Test	Observation
Carbonate (CO_3^{2-})	Add dilute acid.	Carbon dioxide gas given off, which turns lime water milky.
Chloride (Cl^-),	Add a few drops of dilute nitric acid ***then*** add a few drops of silver nitrate solution.	White precipitate.
Sulfate (SO_4^{2-})	Add a few drops of dilute hydrochloric acid ***then*** add a few drops of barium chloride solution.	White precipitate.

b) What is the test for carbon dioxide gas?

KEY POINTS

1 You use a precipitation reaction to identify ions in solution.
2 A precipitate is an insoluble solid. Its colour can help you identify a substance.
3 The sodium hydroxide test (NaOH) uses this idea to identify metal ions.

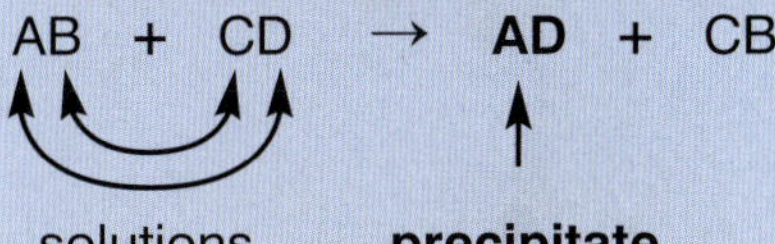

SUMMARY QUESTIONS

1 What is a precipitate?

2 What is the name of the precipitate produced when solutions of copper(II) sulfate and sodium hydroxide react together?

3 What could be the name of the compound dissolved in the following solution?
- When you add sodium hydroxide (NaOH) solution to it, a red-brown precipitate forms.
- When you add dilute nitric acid then silver nitrate solution, a white precipitate forms.

4.8 Flame tests

LEARNING OBJECTIVES

1 What tests can we use to identify chemicals found at a crime scene?
2 How can we improve the accuracy and reliability of the evidence we collect?

Safety

A SOCO recovers a soil sample from the scene of a crime. After a microscopic examination, the forensic scientist decides to find out which metal ions are present in the soil.

The compounds of some metals burn with distinctive colours. So you can detect certain metal ions by doing a flame test. Although a flame test may tell you what metal ion is present, it will not tell you how much of it is in a sample.

Safety: Be aware of the hazards before you try a flame test. Your safety matters. In your risk assessment, consider the risks of hot concentrated hydrochloric acid spitting into your eye or onto the table. See the CLEAPSS Hazcard 47.

a) What do these labels mean?

PRACTICAL

Standard procedure for a flame test

1 Wearing eye protection throughout, clean the nichrome wire loop in concentrated hydrochloric acid.
2 Place the loop in the hot part of a Bunsen flame.
3 Repeat steps 1 and 2 until there is no change in flame colour (other than dull orange). The wire is now clean.
4 Dip the loop in hydrochloric acid again.
5 Dip the loop into the sample. Tap off any crystals that stick to it.
6 Place the loop in the edge of a hot Bunsen flame.
7 Record the colour of the flame.
8 Repeat steps 1 to 4 before testing another sample.

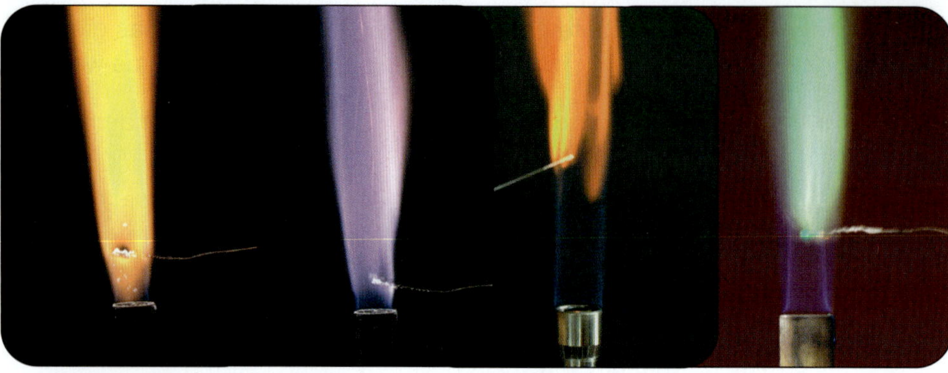

Sodium, Na^{+} Potassium, K^{+} Calcium, Ca^{2+} Copper, Cu^{2+}
Flame test colours of certain metal ions

b) In a flame test, what metal ion gives the following colours?
i) golden yellow ii) lilac iii) brick red iv) blue-green

c) Draw the two tables below. Based on the tests outlined on pages 82–4, identify these two substances found at a crime scene:

Specimen A:

Test	Observation	Conclusion
A flame test was carried out on solid A.	The flame colour was lilac.	
A spatula of solid A was placed in a test tube of hydrochloric acid.	The mixture fizzed. The gas given off turned lime water milky.	
Specimen A is ...		

Specimen B:

Test	Observation	Conclusion
A flame test was carried out on solid B.	The flame colour was blue-green.	
Sodium hydroxide was added to a solution of B.	A blue-green, jelly-like precipitate was formed.	
A spatula of solid B was placed in a test tube of hydrochloric acid.	No reaction.	
A few drops of dilute nitric acid were added to a solution of B. Then silver nitrate solution was added.	A white precipitate formed.	
Specimen B is ...		

DID YOU KNOW?

Dogs have millions more smell receptors than humans. They are trained to detect chemicals such as drugs. The nose knows!

SUMMARY QUESTIONS

1 Copy and complete:

We can use flame tests to find out which . . . . . . ions are in a sample. Different metal . . . . . . produce different flame . . . .

2 Write about a situation when the use of flame tests could prove useful to a forensic scientist.

3 How well do you think a flame test might work on a mixture of salts, containing different metal ions?

4 Your technician gave you 4 mol/dm^3 hydrochloric acid to clean your nichrome wire. It is difficult to remove sodium ions from your wire loop. Your partner suggests using some more concentrated hydrochloric acid. What do you think? Is there any alternative?

KEY POINTS

You can use a flame test to identify a metal ion in a solid:

Metal ion	Flame colour
Sodium	Bright yellow
Potassium	Lilac
Copper	Green-blue
Calcium	Brick red

4.9 Chromatography

LEARNING OBJECTIVES

1. What is chromatography and how does it work?
2. What is the difference between paper and thin layer chromatography?
3. How do we identify the substances present?

Chromatography is a technique for separating and identifying small quantities of chemicals in a mixture. It is ideal for separating out the dyes found in inks. Therefore forensic scientists can use the technique in forgery cases in which paperwork might have been altered.

You have probably carried out a paper chromatography experiment before.

Look at the photos of chromatograms from the ink in a letter and the suspects' pens:
a) Which suspect wrote the letter: A, B, C, D, E or F?
b) Why is this test not a proof of guilt?

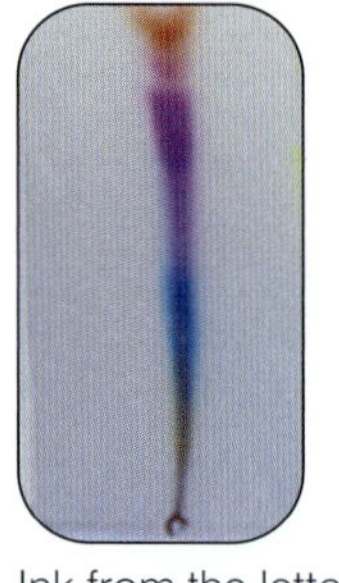

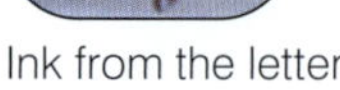

Ink from the letter

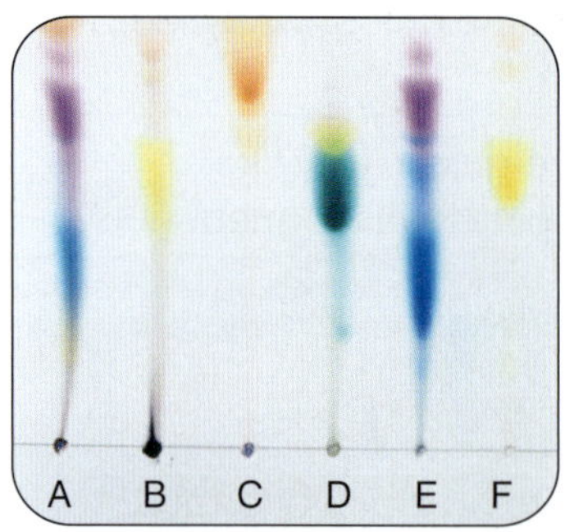

Chromatograms from the suspects' pens

Types of chromatography

There are several types of chromatography. At the cheaper and simpler end there is ***paper chromatography*** and ***thin-layer chromatography*** (TLC). See also DNA profiling and ***electrophoresis*** on pages 88–9.

Every type of chromatography involves a **solvent** passing through a ***stationary substance***. The mixture you are testing dissolves in the solvent. As the solvent moves through the stationary substance, the different chemicals in the mixture separate at different rates.

Some cling more strongly to the stationary substance. These are stopped first. Chemicals that do not bond as easily move on by.

You cannot always use water as the solvent. The mixture may not dissolve in water. Ethanol is a non-aqueous solvent that could dissolve biro ink.

In paper chromatography, the stationary substance is the paper containing trapped water molecules. Paper contains about 10% water.

In thin-layer chromatography, the stationary substance is a tiny layer of powder. The powder is coated onto a plastic sheet or a glass plate. Chemicals moving past attach to the powder at different rates.

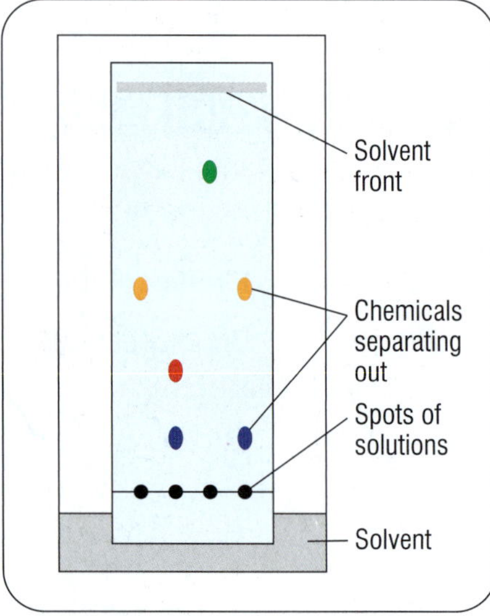

Chromatography

PRACTICAL

Making a chromatogram

1. Dissolve your samples in a suitable solvent. Make sure the solution is concentrated. (If the solvent is flammable ensure there are no naked flames).
2. Place a spot of each sample on a pencil line 2 cm from the bottom of the paper (or TLC plate).
3. Add 1 cm of solvent to your container and stand your paper in the solvent.
4. Cover the container and leave until the solvent rises up ***almost*** to the top.

c) Why are different coloured chemicals in the mixture carried different distances by the solvent?
d) How does a taller container in chromatography help?

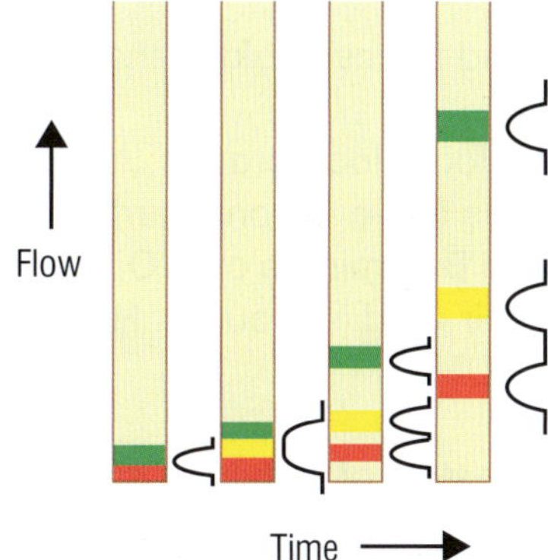

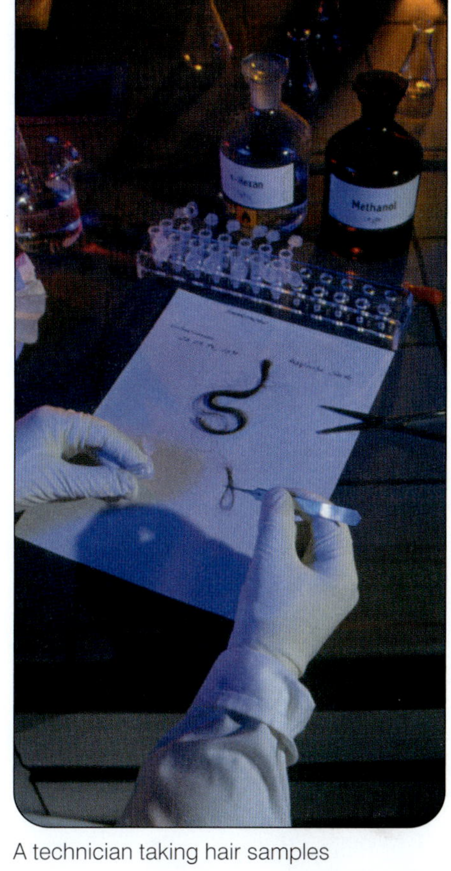
A technician taking hair samples

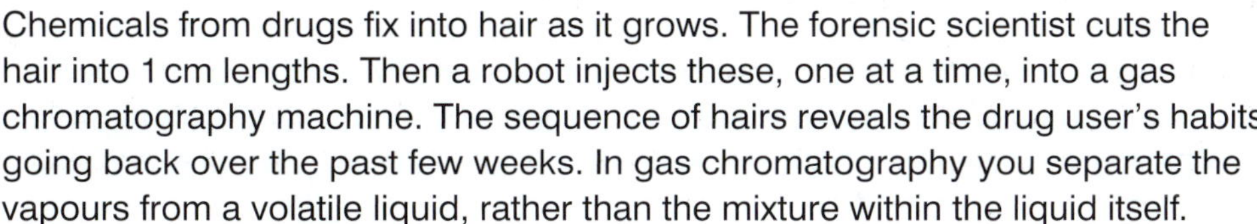

Chemicals from drugs fix into hair as it grows. The forensic scientist cuts the hair into 1 cm lengths. Then a robot injects these, one at a time, into a gas chromatography machine. The sequence of hairs reveals the drug user's habits going back over the past few weeks. In gas chromatography you separate the vapours from a volatile liquid, rather than the mixture within the liquid itself.

SUMMARY QUESTIONS

1. All the substances (A, B, C, D) are pure and not mixtures.
   a) How do we know?
   b) What does the distance *X* measure?
   c) Mark Y is smallest. Is this a problem? Explain your answer.
2. A suspect's hands are covered in paint. How could you match this paint to some graffiti?
3. Why might a forensic scientist decide to view a chromatogram under ultraviolet light?

X
Y
A B C D

KEY POINTS

1. You use chromatography to separate and compare mixtures like inks.
2. Different substances travel different distances, depending on how each clings to the stationary substance.
3. In paper chromatography the stationary substance is the paper containing trapped water molecules. Paper contains about 10% water. In thin-layer chromatography (TLC) the stationary substance is a layer of powder coated onto a sheet.

4.10 Blood and DNA

LEARNING OBJECTIVES

1 What is the composition of blood?
2 Why is DNA useful?
3 What can we conclude from blood tests and DNA profiles?

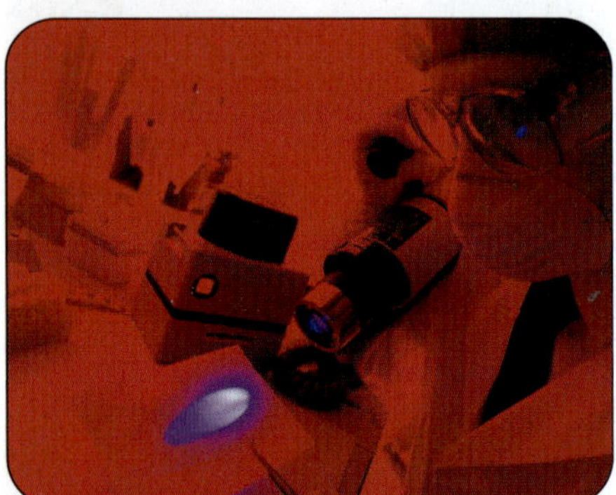
Revealing hidden blood – using a UV lamp to look for signs of blood

You have over five litres of blood circulating round your body!

Blood contains a straw-coloured liquid called plasma. ***Plasma*** carries the red blood cells, white blood cells, platelets and various dissolved chemicals, such as nutrients, round the body:

- ***Red blood cells*** contain **haemoglobin** to carry our oxygen. On the surface of red blood cells are chemicals called ***antigens***.
- ***White blood cells*** defend us against disease. They produce ***antibodies*** to fight bacteria.
- ***Platelets*** help the blood to clot at a wound.

It doesn't fool a forensic scientist if a criminal cleans up a crime scene thinking he will remove the bloodstains. Fluorescin reacts with the iron in haemoglobin, making blood glow under ultraviolet radiation. Spraying with fluorescin and lighting the crime scene with a UV torch reveals the traces of blood.

a) How can a forensic scientist find traces of blood that have been cleaned up?

Most people's blood divides into four blood groups: ***A***, ***B***, ***O*** and ***AB***. The fact that an AB type exists at all implies that everyone carries two genes, one inherited from each parent. A and B dominate over O, so if one parent provided an O and the other an A, then their child is group A. However, A and B do not dominate each other.

b) What can you say about the parents of a child of group AB?

We test for blood groups by adding different antibodies to blood samples. When we add these antibodies, the A or B antigens may, or may not, make the blood clump together. Since it is quick and cheap, forensic scientists may carry out a blood test to eliminate someone from their enquiry. If their suspect and the blood samples from the crime scene do not match, they avoid the expense of a DNA test. Although a DNA test is unique, a single DNA comparison costs £1000!

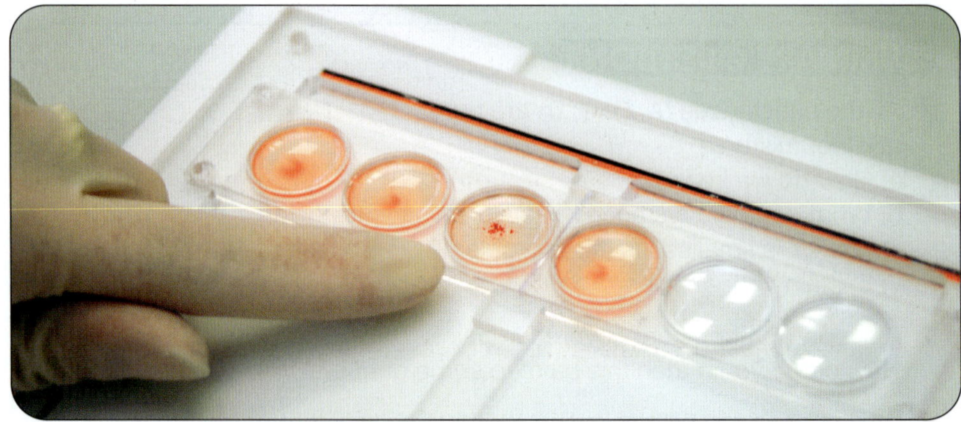
Blood group testing. The sample of blood sticks together if there is a positive test. The presence of antigens on the surface of red blood cells decides the blood group.

Coiled inside the nucleus of the cells of our bodies are metre-long spirals of **DNA**, i.e. deoxyribonucleic acid. Everyone's DNA is different. These long DNA molecules consist of two strands wound around each other in a double helix. They contain our ***genes*** – the genetic traits we inherit from our parents. In fact we get half of our DNA from each parent.

c) Except for identical twins, what can we say about the DNA of every individual?

DNA swab and evidence form. We use a sterile cotton bud to collect cells from the lining of the mouth.

We can extract samples of DNA from blood, saliva, semen and body cells (even hair follicles and dandruff). Forensic scientists analyse saliva from places such as cigarette ends and chewing gum. After sexual assaults they collect semen (from vaginal swabs, underwear and beds), and body cells from under fingernails.

d) What are the advantages and disadvantages of a DNA test compared to a blood test?

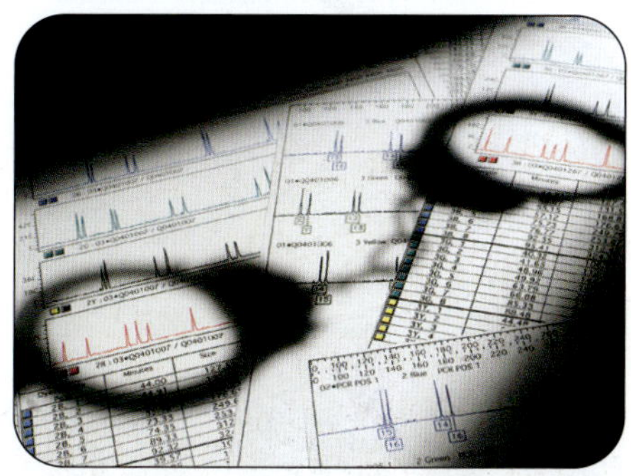

DNA printouts, showing a match between a sample found at the crime scene and one taken from the suspect

To obtain a DNA profile, scientists use an enzyme to cut up the DNA into fragments. Then they separate the fragments by ***electrophoresis***, where an electric field pulls the fragments through a gel.

Different-sized molecules move through the gel at different rates. Think of it as electrically forced chromatography. In electrophoresis, the ***pattern of DNA bands*** produced is ***unique to one individual human being***.

SUMMARY QUESTIONS

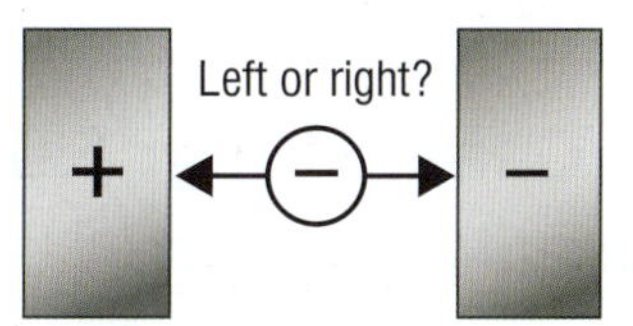

Electric field between plates

1 Look at the diagram. Which way does the negatively charged DNA fragment move?

2 The diagram shows one type of electrophoresis equipment. The gel exerts a frictional resistance on the DNA fragments.

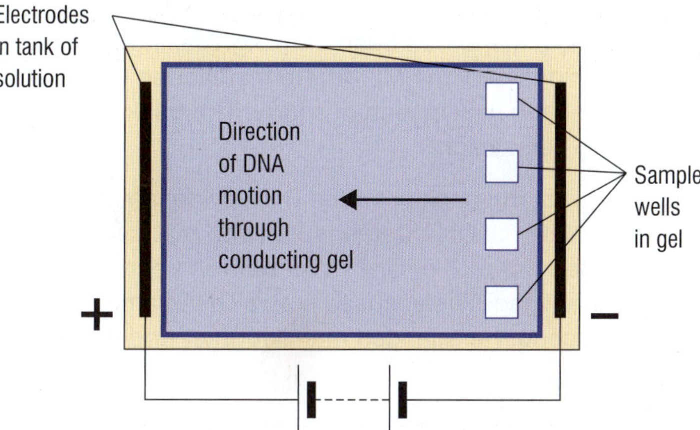

a) Which bits move more slowly through the gel? (Larger bits or smaller bits) Why?
b) Why does their movement separate the fragments?
c) How is this like chromatography?

KEY POINTS

1 Blood contains red blood cells, white blood cells, platelets and plasma. The four main blood groups are A, B, AB and O.
2 Electrophoresis is like electrically forced chromatography. It separates the DNA fragments to produce a DNA profile. Everyone's DNA (like a fingerprint) is unique.

4.11 Databases

LEARNING OBJECTIVES

1. What information do we store on forensic databases?
2. Why are databases searched?
3. How can we record witness descriptions?

Fingerprint and DNA analysis

A database is an electronic filing system. Nowadays national computer databases make fingerprint and DNA comparisons easy.

The ***National Automatic Fingerprint Identification System (NAFIS)*** and the **National DNA Database (NDNAD)** are the key databases used in forensic science.

There are over four million 'criminal justice' profiles on NDNAD. Some of these are DNA profiles from crime scenes. Some are the DNA profiles of convicted offenders.

Other databases are useful in forensic investigations, for example, dental records, insurance company records of valuable items, vehicle records held by the DVLA (Driver and Vehicle Licensing Agency) and police descriptions of missing persons.

a) On which database can you find fingerprint records?

Forensic scientists use databases to record, sort, link and search for information. A match found in a database obviously helps with identification. A mismatch is equally important to eliminate a suspect from a police inquiry.

b) How can information in databases be used in forensic science?

A Case Study: Family DNA link is a crime breakthrough

Using a 'familial DNA search', police successfully prosecuted a man who admitted killing a 53-year-old lorry driver, by throwing a brick through the windscreen of his cab.

The accused man and a friend had been drinking. Making their way home, they crossed a footbridge over the M3, where they hurled bricks at traffic. One smashed through the windscreen of a lorry. The brick hit the lorry driver in the chest, causing heart failure.

Scientific examination of the brick provided DNA evidence. A DNA profile was run against the National DNA Database, but no match was found. Yet 25 people nearly matched the profile. One of these lived nearby and he was a relative of the accused man. The accused man's DNA sample matched the one from the crime scene. He admitted manslaughter and was sentenced to six years in prison.

A crime victim works with an identikit artist

Methods to record the description given by a witness include '***artist's impression***' and '***identikit***' reconstruction. Identikit reconstruction involves using standard face shapes, hair, skin, eye colours and other features such as glasses. The artist modifies these, with the help of the victim, until the appearance is recognisable. The public can then help the police to identify possible suspects.

A police artist's impression of a criminal

SUMMARY QUESTIONS

1 How can DNA analysis solve a crime when there is no suspect?

2 "The innocent have nothing to fear from a DNA database of the whole population."

"A UK-wide database erodes my civil liberties."

What do you think?

3 The computer print outs of the DNA profiles A and B, shown in the diagram, are a good match. What conclusion can you make?

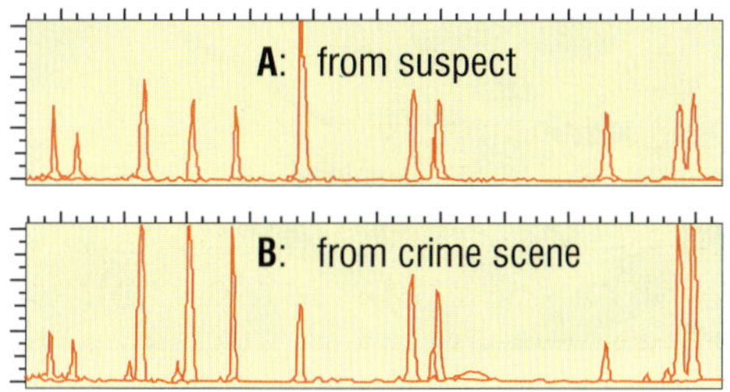

Computer profile printouts of DNA, A–B

4 When checking the National DNA Database (NDNAD), what does it prove for each of the following?
   a) The DNA profile of a suspect matches that of a person on NDNAD.
   b) The DNA profile from a crime scene matches that of a person on NDNAD.
   c) The DNA profile of a suspect matches the DNA profile from a crime scene.
   d) The DNA profile from one crime scene matches the profile from another crime scene.

5 Look at the DNA profiles, C–E. What conclusion can you make?

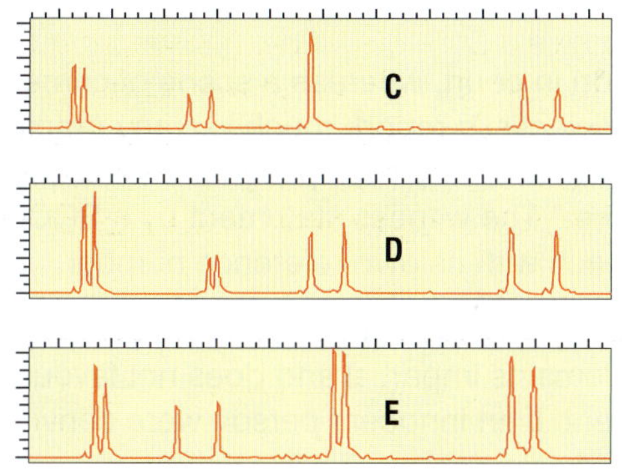

Computer profile printouts of DNA, C–E

KEY POINTS

1 A database match can link a suspect to a crime. A mismatch can exclude a suspect from an investigation. Examples include NAFIS (National Automatic Fingerprint Identification System) and NDNAD (National DNA Database).

2 The descriptions of a criminal by a witness may be recorded using 'identikit' or an 'artist's impression'.

4.12 Presenting evidence

LEARNING OBJECTIVE

1 How do forensic scientists present evidence to a court?

After collecting and labelling evidence from the crime scene and all the laboratory investigations, comes the court case. The forensic scientist's report needs to stand up in court.

The purpose of the report is to summarise what you have done and what you have found out. For any forensic report, all exhibits must be included, whether they prove the involvement of a suspect or not.

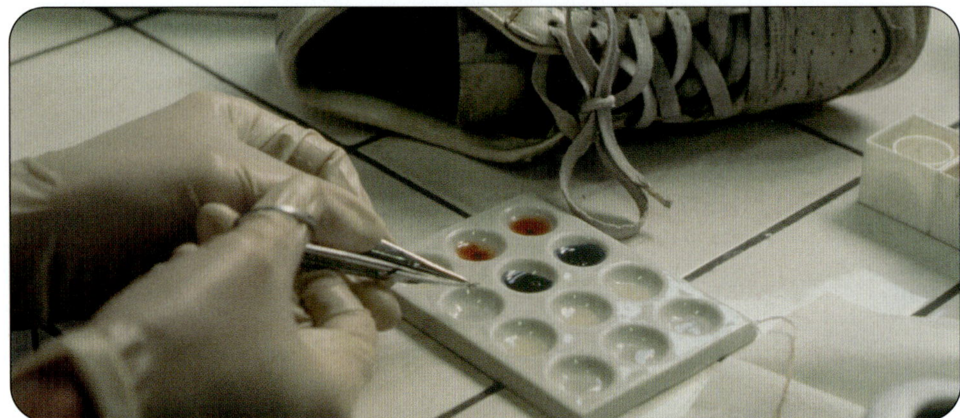

Forensic test of a blood sample from a shoe

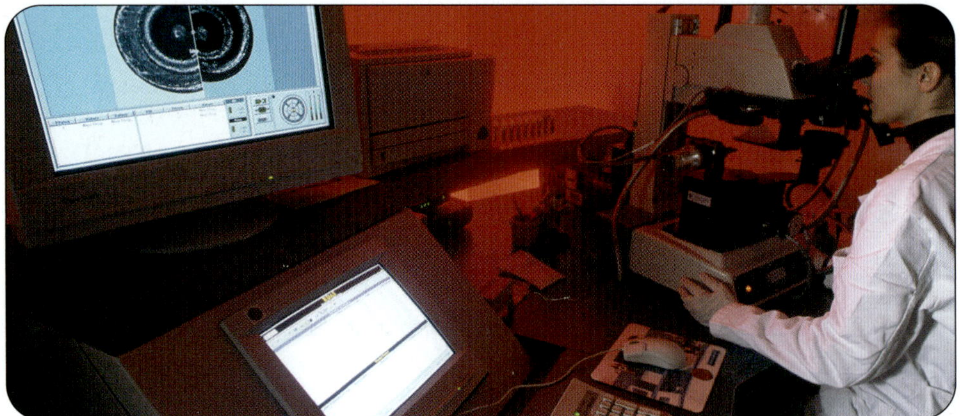

Comparing bullet cartridges. When it fires, a gun makes a distinctive mark on the bullet.

A forensic scientist can give an opinion in court, whereas a scene of crime officer (SOCO) cannot. The forensic scientist's report, much like any practical report, ends with a conclusion. For example, "It is my opinion that the liquid was not responsible for starting the fire." The witness statement by a SOCO merely lists the evidence collected, each with its own reference number.

The defence, as well as the prosecution, can ask a forensic scientist to write a report. It is important that the report remains impartial and does not favour any one side. The suspect may be innocent. If an innocent person were convicted, there would be a miscarriage of justice.

Here is an example of a forensic scientist's report:

Forensic Science Service (FSS)

Case against: *Gary Otya (d.o.b.: 20/06/1983)* Lab Ref. No. *FSS1*

Statement of: *Brian Smart BSc*

Age of witness: *Over 21* Occupation: *Forensic scientist*

This statement is true to the best of my knowledge and belief and I make it knowing that, if it is tendered in evidence, I shall be liable to prosecution if I have wilfully stated in it anything that I know to be false or do not believe to be true.

Dated: *1st May 2007* Signature: *B Smart*

Qualifications and experience: *I am a BSc graduate. I have been employed by the FSS since 1999, specialising in the analysis of body fluids, alcohol and drugs.*

Receipt of exhibits: *On the 1st April 2007 blood labelled 1S1 was taken from Gary Otya for examination.*

Circumstances: *Mr Otya was arrested on suspicion of driving whilst under the influence of alcohol or drugs following an incident at 00:30 on 1st April 2007. I was asked to analyse the above blood sample for the presence of any drugs that might have impaired Mr Otya's ability to drive.*

Results: *The blood sample was analysed for a range of drugs that can have a detrimental effect on the ability to drive safely. SOPs (Standard Operating Procedures) were used. Methadone was found in the blood (0.30 milligrams per litre). No alcohol and no other drugs were found in Mr Otya's blood sample.*

Conclusions: *The quantity of methadone in the blood sample implies that Mr Otya consumed methadone at some time immediately before the arrest. Methadone is prescribed to treat dependence on heroin. The concentration of methadone found in Mr Otya's blood is consistent with normal medical use, and illicit abuse, of this drug. Side effects of taking methadone include nausea, vomiting, drowsiness and confusion, which can significantly affect driving performance.*

To summarise: *Methadone was detected in Mr Otya's blood sample. Methadone causes effects that can undermine the ability to drive safely.*

Signed: *B Smart* Dated: *1st May 2007*

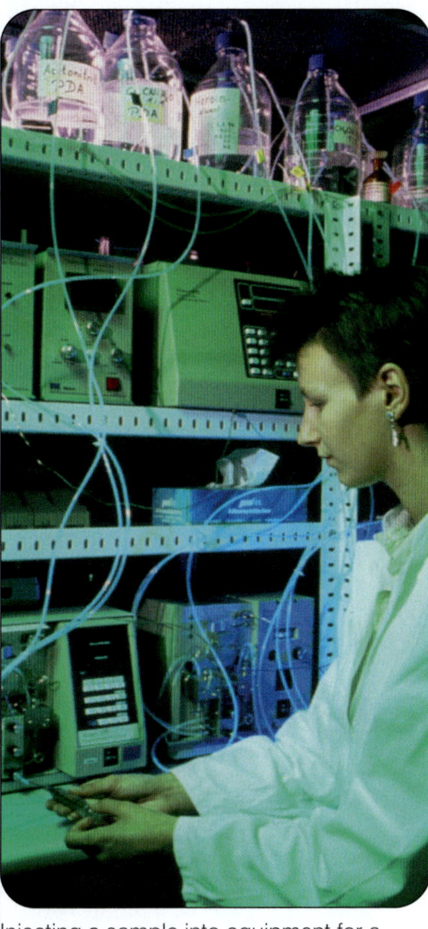

Injecting a sample into equipment for a forensic drugs test

a) What is the case about in the forensic scientist's report?
b) What was the forensic scientist asked to do?
c) What are the jury likely to imply from Brian Smart's report?

SUMMARY QUESTIONS

1 What are the differences between a SOCO witness statement and a forensic scientist's report?

2 Why must the same reference number be used from the time evidence is seized to the time that evidence is presented to the court?

KEY POINTS

1 A forensic scientist's report presents evidence and conclusions for a court.

2 Conclusions, based on the evidence, state whether or not a suspect was at the crime scene or committed the crime.

4.13 "Holmes, that's incredible, how did you know that?"

Sherlock Holmes was a detective in novels written in Victorian times. He was brilliant at solving crimes from clues left by criminals.

Nowadays, finding the evidence and figuring out what it means is the work of a forensic scientist.

a) Hair, fibres, insects, fingerprints, handwriting, and teeth! Explosives, bite marks, bullets, and blood! What's the link between all these things?

Who committed the crime? When? Was she drinking? Was he using drugs? Is this really her signature? How long has he been dead? What was the cause? How fast were they going? Is this the gun? These kinds of questions come up in crimes like murders, drug cases, and burglaries.

Scene of crime officers (SOCOs) entering a crime scene

Finding out when somebody died, and how a factory or an office was broken in to, and what caused an accident are just a few of the everyday mysteries that need to be investigated.

There are many kinds of forensic science jobs. Scene of crime officers (SOCOs) go out to find evidence on the spot. Others, like toxicologists (experts in harmful substances), concentrate on lab work. A ballistics expert deals with firearms and bullets. An entomologist knows all about insects and how to use them in solving mysteries, such as finding out when somebody died.

Detective Sherlock Holmes on the case!

Using fingerprints to identify people was a big step forward in Sherlock Holmes' day. Nowadays, we also use 'DNA profiling' to identify people.

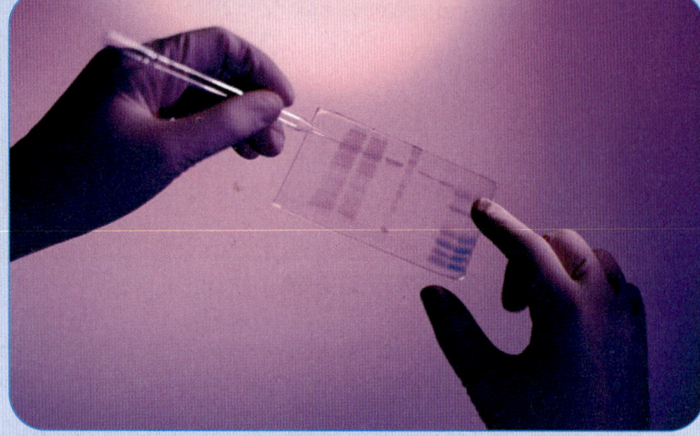

DNA profiling is based on each person's own unique genetic code

COURSEWORK HINTS

Today, science plays a key part in policing. Why not contact your local police force and arrange to meet someone from 'scientific services' or 'specialist units'? See what information is available on-line at www.**mycounty**.police.uk (substitute suffolk, warwickshire, etc. for '***mycounty***'). This will support your Unit 3 coursework (stages 3A and 2E/3E) to reseach and explain the vocational significance of your investigation.

Would forensic science be right for you?

To test your interest, read a few mysteries and play some crime-solving computer games. Forensics can be fascinating, but it also involves careful thinking and precise testing. The skills you need come from studying science.

CRIME SCENE DO NOT ENTER CRIME SCENE DO NOT ENTER

Murder most foul

– reconstruction of the crime scene

Outside the kitchen:

Underneath the broken window, the scene of crime officer (SOCO) sees a shoe print in the flower bed. She makes a cast of this and puts glass fragments into bags, for future comparison. There is a smear of blood and some clothing fibres on the broken pane. After collecting these, the officer dusts the window sill and pane for fingerprints.

Inside the kitchen:

Some more fingerprints are on the inside of the frame. On the work top there is a faint shoe print. The SOCO takes an impression of these using a gelatin lift. There is a cigarette stubbed out on the worktop. The officer keeps it to get a DNA profile from the saliva on the butt.

Inside the bedroom:

On the floor by the door is a crumpled note, written to a man and signed by the victim. This is sent off for document analysis. On the walls are spots of blood. The SOCO photographs the stains and records their size and splatter pattern. She takes a sample for DNA analysis. She also dusts for fingerprints on the door and around the room. The bedding, the victim's clothing and the body are 'taped' to recover any hairs or fibres. Finally, at the post mortem, body fluid samples are taken from the victim for toxicological testing.

b) How many pieces of evidence has the SOCO collected?

Outside the kitchen

Inside the kitchen

Inside the bedroom

SUMMARY QUESTIONS

1 Complete the sentences using the words below. Words may be used more than once.

arch 'artist's impression' clothing contaminating dental eliminate fingerprints link loop recording restricting scene scene of crime officers vehicle whorl

SOCOs (. . . . . .) and forensic scientists avoid . . . . . . evidence at the crime . . . . . . by . . . . . . access, wearing protective . . . . . . and using appropriate methods of sampling, storage and . . . .

The three distinctive types of fingerprint pattern are . . . . . ., ... and . . . .

The police record the description of a witness using . . . . . . and 'identikit'. Information stored on databases includes DNA, ..., ... records, and . . . . . . records (held by the DVLA). A mismatch is useful to . . . . . . a suspect from a police enquiry. Forensic scientists use databases to record, sort, ... (or match) and search for information.

2 Complete the sentences using the words below. Words may be used more than once.

$CaCO_3$ carbon dioxide copper hydroxide covalent $CuCl_2$ energy ethanol giant glucose high KCl low melting point Na_2SO_4 negative organic positive separate strong water weak

The structure of ionic compounds is a . . . . . . lattice, which is held together by the attraction between . . . . . . and . . . . . . ions. Ionic compounds have a high . . . . . . because we need a lot of . . . . . . to break the bonds between the ions.

What is the formula of each of these?

potassium chloride. . . . . ., calcium carbonate . . . . . ., sodium sulfate. . . . . ., copper(II) chloride . . . .

Substances obtained from living materials are . . . compounds with . . . . . . bonding. CO_2 is . . . . . ., H_2O is . . . . . ., C_2H_5OH is . . . . . . and $C_6H_{12}O_6$ . . . . The covalent bonds between the atoms in a molecule are . . . . Covalent compounds have . . . . . . melting points because the forces between the molecules are . . . . . . and we only need a little energy to . . . . . . them.

The precipitate that forms when copper(II) sulfate reacts with sodium hydroxide is . . . .

3 Complete the sentences using the words below. Words may be used more than once.

A AB B disappears immersing match mixture normal O plasma platelets red blood cells refractive index refracts solvent stationary strongly white blood cells

In chromatography some colours in the . . . . . . don't move as far in the . . . . This is because they cling more . . . . . . to the . . . . . . substance they pass through.

Blood contains . . . . . ., ..., ... and . . . . The four main blood groups are . . . . . ., ..., ... and . . . .

When a ray of light enters glass it . . . . . . towards the . . . . We determine the . . . . . . of glass fragments by . . . . . . them in oil. When the values for the glass and oil . . . . . . the glass . . . .

KEY PRACTICAL QUESTIONS

1 Describe how to:

a) Collect a fingerprint from a glass surface.

b) Collect a sample of dry blood.

c) Make an impression of the sole of a suspect's shoe.

d) Carry out a flame test.

e) Obtain a clear solution and test for the solubility of a substance dissolved in that solution.

2 Why do we use these?

a) universal indicator paper

b) a flame test

c) lime water

d) acidified potassium dichromate

e) Benedict's solution

f) chromatography

g) electrophoresis.

EXAM-STYLE QUESTIONS

1 a) State what the letters 'SOCO' mean. (1)
b) Give a reason why a SOCO would make a plaster cast. (1)
c) Describe why a SOCO might hold a bright light at an oblique angle. (1)
d) State why a SOCO puts a scale beside evidence before taking photographs. (1)
e) Apart from photography, suggest how a SoCO could reveal a fingerprint. (1)
f) SOCOs are civilians, who work with the police. Describe why a SOCO must remain impartial. (2)

2 Shoe prints are often found at the scene of a crime. The photograph shows a shoeprint left at a crime scene in an area of damp soil.

a) Give ***two*** reasons why it is important for the forensic scientist to have first access to a crime scene. (2)
b) Describe how to make a plaster cast of a shoe print found in an area of damp soil. (3)
c) In addition to making the cast, what other method could be used at the crime scene to make a permanent record of the shoe print? (1)
d) Features from the cast are used to match the footprint with a suspect's shoes. Give ***three*** features from the cast that might be used in this way. (3)
e) A sample of soil was removed from the flowerbed where the cast was made. A forensic technician was given the job of testing the soil.

Method:
1 Shake a weighed sample of soil with 100 cm^3 of distilled water.
2 Filter the sample using filter paper and a funnel.
3 Test the pH of the filtrate.

Results:
Mass of soil and container = 11.36 g
Mass of container = 7.15 g
pH of filtrate = 6.50

i) Calculate the mass of soil used in the test. (2)
ii) Complete the following sentences using words from the list:
alkaline clear insoluble neutral soluble
The mixture is filtered to remove the . . . soil particles.
After filtering, the liquid is . . .. (2)
iii) How would the technician test the pH of the filtrate? (1)
iv) Suggest how tests on the soil could be useful at a later stage in the investigation. (2)

3 a) Use the diagram to explain the statement: "Every contact leaves a trace."

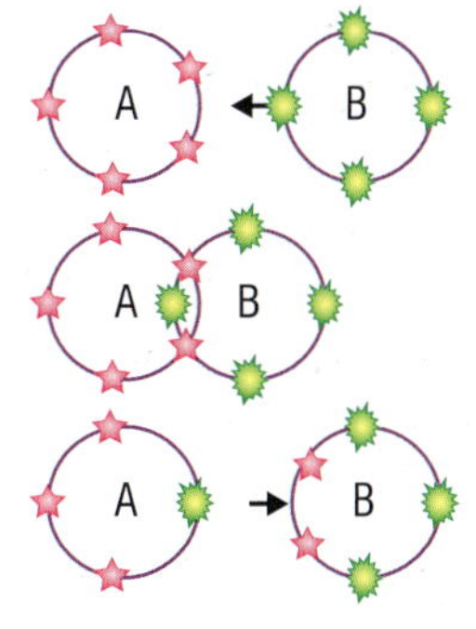

(5)

b)

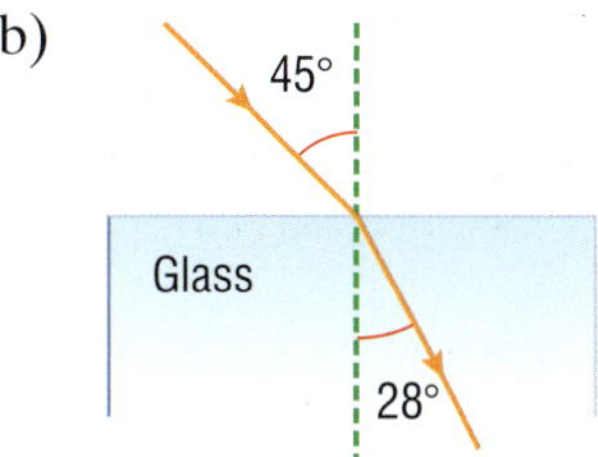

i) What happens to light at a glass surface? (2)
ii) Write the formula for refractive index. (1)
iii) Using the values shown in the diagram, calculate the refractive index of the glass. (2)

c) Describe how to determine the refractive index of a glass fragment by the oil immersion method. (3)

EXAMINATION STYLE QUESTIONS

1 (a) Label the fingerprints correctly with the words *loop*, *arch* or *whorl*. (*2 marks*)

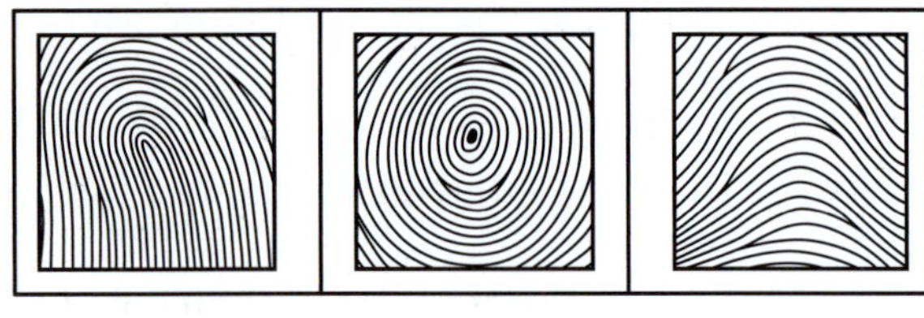

(i) (ii) (iii)

(b) State one piece of information that is put on a fingerprint identification card. (*1 mark*)

(c) (i) Name two other types of impressions left at crime scenes. (*2 marks*)
(ii) Describe a method to make a permanent record of one of these impressions. (*2 marks*)

(d) For what purpose would an artist's impression be made? (*1 mark*)

(e) Fingerprints are stored on a database called the National Automatic Fingerprint Identification System (NAFIS).
(i) What is a database? (*1 mark*)
(ii) Name another database used to store records. (*1 mark*)

2 (a) Following a reported stabbing of a young adult, a crime scene investigator finds three blood splashes on a path as shown below. Where exactly did this stabbing take place, A, B, C or D?

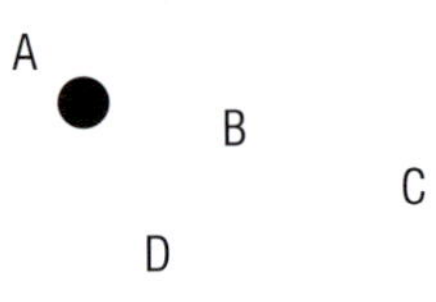

(*1 mark*)

(b) A young man, Lenny Little, attends a nearby hospital with a blood soaked handkerchief round his arm, but denies being stabbed. His blood is tested for blood grouping. A blood test is also carried out on the blood recovered from the path where the stabbing took place.
(i) Name the four common blood groups. (*3 marks*)
(ii) Suppose the two blood tests show that the blood groups do not match. State what this proves. (*1 mark*)
(iii) However, the blood groups do match and Lenny confesses to being the victim of the stabbing by a Mr Barry Big. Suggest a reason why Lenny denied being stabbed originally. (*1 mark*)

GET IT RIGHT!

You make plaster casts and take photographs of the impression at the crime scene. You ink the footwear or tyre after recovering it from the suspect. You **do not** make plaster casts of the footwear or tyre.

GET IT RIGHT!

The number of marks per question is a guide to the detail you must write. If there are 2 marks you must make two points. **'Explain'** means 'give reasons for' and expects a more detailed answer.

(c) The police find Barry at his home with blood on his clothes. A DNA test of the blood on Barry's clothes matches that of Lenny's DNA.
   (i) Explain why DNA tests were carried out rather than a blood test? (*2 marks*)
   (ii) The DNA matches that of a sample on the National DNA Database. Explain the information this provides to the police. (*2 marks*)

3 (a) Some trace evidence was recovered from a crime scene. Give two examples of trace evidence. (*2 marks*)

(b) The evidence was viewed through a polarising microscope. State why a polarising microscope is useful. (*1 mark*)

(c) Among the evidence were some tiny crystals. A flame test was carried out on these crystals. Why was the flame test carried out? (*2 marks*)

(d) A few of the crystals were dissolved in water and a sodium hydroxide test was performed. This confirmed the result of the flame test. In the sodium hydroxide test what do you hope to observe. (*2 marks*)

(e) Further tests proved the chemical was the ionic compound copper nitrate.
   (i) What is an ion? (*1 mark*)
   (ii) What is the formula for copper nitrate? (*2 marks*)

GET IT RIGHT!

Practise writing the formulae for simple ionic compounds.

Remember:

_ '-ides' like sulf*ide*, S, have just the element.

_ '-ates' like sulf*ate*, SO_4, also contain oxygen.

4.14 Forensic science coursework tasks

LEARNING OBJECTIVES

1 What is involved in my Unit 3 coursework investigation task for forensic science?
2 How do I maximise my grades for this coursework?

Searching for clues in the wreckage of a burnt-out house. Evidence of how the fire started may determine whether it was an accident or arson.

Forensic scientists work closely with the police and are sometimes required to go to a crime scene, such as a murder or fire. They aim to provide impartial, scientific evidence for use in courts of law.

Carrying out more than one Unit 3 investigation will help you to perform better in the Unit 2 exam, and help you progress through the Unit 3 stages of this course. Yet remember, for your coursework, you only need to produce a report of one practical investigation set in a vocational context. See pages 156–7 for the coursework checklist and mark scheme. 40% of your GCSE marks come from this one report. You should carry out more than one coursework experiment. This will include a number of tests and techniques for comparing and matching samples. The aim is to indicate whether or not we can link a 'suspect' to a crime.

a) Why does a match not necessarily prove that the suspect did commit the crime? Give an example in your answer.

In the report of your investigation:

- Describe the purpose of your tests. Explain the vocational importance of these tests.
- If appropriate, say how you are collecting and preparing your samples. Include a plan and a risk assessment for your tests.
- Evidence could include photographs or drawings from your 'crime scene', as well as experimental evidence.
- Make conclusions from your evidence and evaluate the complete investigation. Use the facts to explain and justify your conclusions. Include any evidence that does not fit with your conclusions.
- Explain how a forensic scientist might use the results of your investigation to indicate the probability of a 'suspect' being linked to the crime.

The two suggested forensic science investigation tasks are:

- The case of the lost laptop (pages 106–7): revealing and examining fingerprints on different surfaces. You produce plaster casts of footprints and measure tyre tracks. You also analyse bloodstains.
- The case of the vindictive vandal (pages 108–11): using microscopy to examine surface details and cross-sections of paint layers. You compare samples, e.g. types of clothing, fibre, hair or small seeds. Then you nail your suspect using chemical reactions and flame tests to detect for the presence of Na^+, K^+, Ca^{2+}, Cu^{2+}, Fe^{2+}, Fe^{3+}, Pb^{2+}, Cl^-, SO_4^{2-} or CO_3^{2-} ions.

The following two tasks are not detailed enough for a Unit 3 coursework investigation, but could be used as mock investigations to develop your practical skills:

- The case of the fruit salad (pages 102–3): identifying the contents of a liquid and measuring the concentration and solubility of the substances dissolved in the solvent.
- Testing for ethanol (pages 104–5): carrying out a higher level titration to test for ethanol as in the original breathalyser.

A driver, suspected of drink-driving, breathing into the mouthpiece of a breathalyser. Even if he tests negative, he may still need to take a drugs test. Sadly, driving under the influence of drugs is an increasing hazard. 60% of drivers and passengers in fatal road crashes have drugs in their blood.

SUMMARY QUESTIONS

1 What is another way of writing, 'Explain the vocational importance of these tests.'?

2 a) What is a 'solvent'?

b) Name an 'organic solvent'.

3 What do the following symbols stand for?

Na^{+}, K^{+}, Ca^{2+}, Cu^{2+}, Fe^{2+}, Fe^{3+}, Pb^{2+}, Cl^{-}, SO_4^{2-}, CO_3^{2-}

4 Research the Road Traffic Offenders Act 1988. You probably know that the legal blood alcohol consumption (BAC) limit in the UK is 80 mg of alcohol per 100 ml of blood (80 mg/ml). Find out the limit in micrograms of alcohol per 100 ml of breath (i.e. μg/ml).

4.15 The case of the fruit salad

LEARNING OBJECTIVES

1. How do you measure pH?
2. How do you obtain a clear solution?
3. How do you measure concentration?
4. How do you test for solubility?

In this mock investigation you need to be able to describe how to measure the pH and concentration of a solution. You also need to know how to obtain a clear solution and how to test for solubility. Look at the following case study. Then carry out the practical.

COURSEWORK HINTS

This task is designed to help you to improve your practical skills before you embark on a coursework task.

Did someone tamper with the fruit salad?

It was the night of the Applied Science party in the Year 11 common room. Sheila came with me. A few minutes after we arrived, Sheila ate some fruit salad. With the first mouthful of fruit salad came the scream, “Water!” She spat it out. Her drink of water probably saved her. This was not like her at all. Our teacher, Mr Jones, came across to see what all the noise was about. “There’s something in this fruit salad, Sir. It burnt my mouth.”

Mine had seemed alright and Mr Jones had finished his. “Maybe a bit too much lemon if it was a bit sharp” remarked Mr Jones. “No”, insisted Sheila, “there’s something really wrong.”

Out of his pocket appeared a strip of universal indicator paper. “How weird is this guy?” I thought. But when Mr Jones put it in the fruit salad . . . it turned ***red***.

The police didn’t believe Mr Jones when we phoned, saying “Too much to drink.” “Teenagers having a laugh.”

“Right, you two, some forensic science. Bring Sheila’s fruit salad and your own down to my lab.” After our analysis Mr Jones phoned the police again. They believed him this time and got to the school before we had time to walk back to the common room.

acidic 2 4 ph 7 alkaline 9 11

Universal indicator paper

So what happened? What did we do in the lab? And who was to blame? I can’t answer the last question. The case is ongoing, but why don’t you do what we did that night? Your teacher has the details and can simulate the contents of the two fruit salads for you to test.

a) Why is the boyfriend’s fruit salad needed as well as Sheila’s? Use the word ‘control’ in your answer.

Perform the same tests on each sample of fruit salad, see below. (**Safety:** Realise that Sheila’s fruit salad is at least an irritant. Write a risk assessment before you start. Remember you are not allowed to taste anything in a laboratory.)

PRACTICAL

Testing the fruit salad

Wearing eye protection use the following steps to guide you through the analysis:

1 **Filter** the fruit salad to obtain a clear solution. Test its **pH** with universal indicator paper.

2 The two acids that students commonly have access to are hydrochloric acid and sulfuric acid. To find out which one is in the fruit salad, put a little of the solution into a test tube with a teat pipette. First add a little dilute nitric acid. Then add an equal quantity of silver nitrate solution. If the solution turns white the acid is hydrochloric acid (and not sulfuric acid).

- What is the acid in Sheila's fruit salad?

See the CLEAPSS Hazcard for hazards associated with this acid.

3 Use a syringe to collect 20 cm^3 of the solution.

- Why don't you need the needle?

4 Put your 20 cm^3 of solution into a conical flask. Add a few drops of screened methyl orange indicator, this turns pink to green in an alkali. Fill a burette with 0.4 mol/dm^3 sodium hydroxide. (**Safety:** This alkali is also an irritant.) ***Slowly*** add the alkali to your solution. Measure the volume of alkali you have to add to make the solution neutral – when the indicator just changes colour. Call this volume V cm^3.

- If you need 20 cm^3 of alkali to neutralise your solution, it must be a 0.4 mol/dm^3 concentration. Why?
- You know the volume of alkali you added to reach the end point, so now calculate the concentration of the acid in the solution. Compare this with the concentration of the boyfriend's control sample.

6 Weigh an evaporating dish. Mix 20 cm^3 of fresh solution with V cm^3 of alkali in an evaporating dish. Place this on a beaker of boiling water over a Bunsen flame.

- Why don't you heat the solution directly while you evaporate it away?
- Stop heating when the solution goes tacky. What does it look like now?

6 Allow to cool, then scrape away the salt from the rim. Reweigh the evaporating dish and its sugary contents. Calculate the mass of the sugary residue. Your answer is the mass of sugar in 20 cm^3 (or 20 g) of fruit salad solution.

- So what is the solubility of the sugar in both samples in units of grams of sugar per 100 g of water?
- When Sheila put the fruit salad in her mouth, the criminal was hoping to mask the taste of the acid. How?
- What conclusions can you make based on your evidence?

Titrating the solution. Stop adding sodium hydroxide when you reach the 'end point', when the solution just turns to neutral.

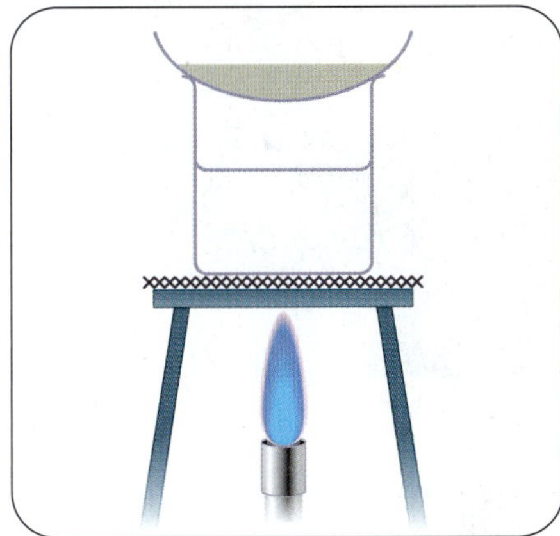

Evaporating the solution

KEY POINTS

1 Use universal indicator to measure pH.
2 Filter to obtain a clear solution.
3 Titrate to measure concentration.
4 Evaporate and weigh residue to test for solubility.

4.16 Testing for ethanol

LEARNING OBJECTIVE

1 How do you test for ethanol?

The alcohol in alcoholic drinks is ethanol, C_2H_5OH (or $CH_3–CH_2–OH$). Breathalysers use a chemical reaction involving ethanol that produces a colour change. Photocells connected to a meter then monitor that colour change. The test for ethanol in the original breathalyser used acidified potassium dichromate, which changes from yellow to green.

Alcohol test. You test for ethanol with acidified potassium dichromate, which changes from orange to green.

The only safe way is not to ***drink*** alcohol at all if you are ***driving***.

These are the legal limits for driving:

- Breath 35 mg/100 cm^3.
- Blood 80 mg/100 cm^3.
- Urine 107 mg/100 cm^3.

Over 7000 UK drivers failed breath tests last year.

Over 3000 people were killed on the roads and more than 30 000 were seriously injured.

One in six of all deaths involve drivers who are over the legal alcohol limit.

The most dangerous times of day are when children go to school and people return from work.

The breath test

Young men aged 17–29 are the main casualties and drink-drive offenders.

If you're convicted of drink-driving you will:

- Lose your licence for at least 12 months.
- Face a maximum fine of £5000.
- Face up to six months in prison.
- Pay up to three times as much for car insurance.

Carry out the following practical and procedure for testing ethanol as a mock investigation:

PRACTICAL

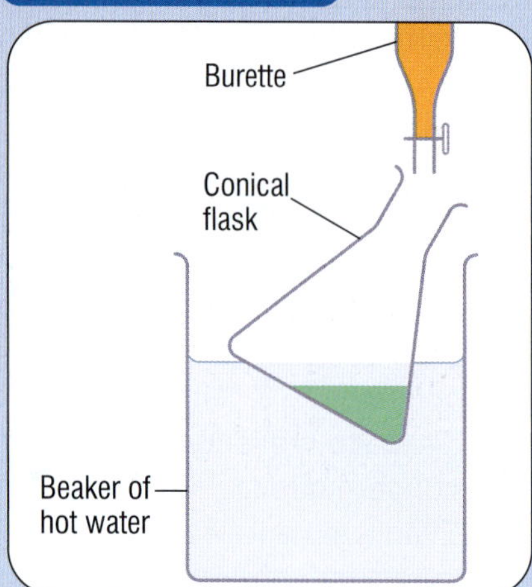

Measuring the alcohol content of white wine by titration

The reaction is slow, so heat the reactants in a water bath at 50 °C.
Safety: Do not heat the ethanol directly. See the CLEAPSS Hazcards 40, 78 and 87 and write a risk assessment before you start. Wear eye protection.

Standard procedure:

1 Using a pipette transfer 10 cm^3 of white wine into a conical flask. Add a few drops of silver nitrate solution to enhance the colour at the end point. Heat this in a beaker of water from a kettle. Aim to maintain the temperature at 50 °C during the titration.

PRACTICAL

2 Fill a burette using a funnel with your ***standard solution*** of acidified potassium dichromate. The potassium dichromate is 0.5 mol/dm^3 (0.5 moles $K_2Cr_2O_7$ were diluted in 1 dm^3). Drain some of the liquid from the burette into a beaker to remove the air below the tap. Record the starting volume to the nearest 0.05 cm^3.

3 Slowly add the dichromate (1 cm^3 at a time) to the flask of white wine. Swirl the flask and stop at the end point, when a red-orange colour first appears. Record the final volume. Calculate the volume of acid used. This is your trial or rough experiment.

4 Repeat the titration more carefully, swirling and adding acid slowly drop by drop near the end point. Continue repeats until you get two accurate volumes within 0.1 cm^3. Average the two accurate titration results.

The equation for the reaction, oxidising alcohol, is complex:

ethanol + sulfuric acid + potassium dichromate ⟶ ethanal + potassium sulfate + chromium sulfate + water

$$3CH_3CH_2OH + 4H_2SO_4 + \underset{(orange)}{K_2Cr_2O_7} \longrightarrow 3CH_3CHO + K_2SO_4 + \underset{(green)}{Cr_2(SO_4)_3} + 7H_2O$$

Assume we use an average potassium dichromate volume = 22.0 cm^3.

To calculate the concentration and alcohol content of the wine:

1 See that the balanced chemical equation shows that: 3 moles of CH_3CH_2OH react with 1 mole of $K_2Cr_2O_7$

	$3CH_3CH_2OH$	$1K_2Cr_2O_7$
Volumes:	10 cm^3	22.0 cm^3
Concentrations:	?***M mol/dm^3*** (? = unknown concentration)	0.5 mol/dm^3

2 Now use this ***formula*** to work out the unknown concentration:

$$\text{(ethanol)}\quad \frac{\textbf{concentration} \times \textbf{volume}}{\textbf{number of moles}} = \frac{\textbf{concentration} \times \textbf{volume}}{\textbf{number of moles}}\quad (K_2Cr_2O_7)$$

$$\frac{M \times 10}{3} = \frac{0.5 \times 22.0}{1}$$

Therefore

$$M = \frac{0.5 \times 22.0 \times 3}{1 \times 10}$$

$$M = 3.3\ \text{mol/dm}^3$$

3 The relative formula mass of ethanol, CH_3CH_2OH = (2 × 12) + (6 × 1) + 16
= 46

1 litre = 1 dm^3 or one decimetre cubed. The mass of 1 litre of water is equal to 1000 g.
So the mass of ethanol in the one litre bottle of wine = 3.3 × 46 g
= 152 g

A 1 litre bottle of white wine is mainly water, so the alcohol content of the wine = (152 g / 1000 g) × 100%
= 15.2%

Repeat the steps of your calculation with the average volume of acid that you used in your titration.

4.17 The case of the lost laptop

LEARNING OBJECTIVE

1 How do you investigate print evidence?

In this coursework task, imagine you are a crime scene investigator who is asked to collect evidence from the scene of a crime, and then to see whether there is a link with any of three possible suspects.

Who stole the laptop?

One of the sixth form students at Smellybridge High School has had their laptop stolen. It looks as though the burglar entered through a window and in doing so, cut his/her hand. Forensic scientists have already taken away blood samples and glass fragments.

Your job, as a crime scene investigator (CSI), is to photograph, retrieve and analyse other evidence.

Three areas look promising, as there are:

- Fingerprints found where the burglar entered.
- Several reddish stains on the floor which look like blood.
- Some footprints found outside the broken windows of the building.

a

b

c

d

Arrowheads

e

Examples of the bloodstains on the floor

The reddish stains on the floor appear to be blood splats. Splats d and e, furthest from the table have ***arrowheads***. The arrowheads point away from the table where the laptop was. Maybe the burglar shook blood off his/her hand before picking up the laptop. You assume that drops landing at a smaller angle have longer arrowheads.

a) How could you test your assumption about the length of arrowheads back in the laboratory (without using real blood)?

Planning your coursework

Think about the evidence that needs to be collected and break down your three tasks:

- Revealing, photographing, lifting and storing the fingerprints, see page 73.
- Photographing and making plaster casts of the footprint, see page 78.
- Testing the theory: the shallower the angle that drips of blood land, the longer their arrowheads.

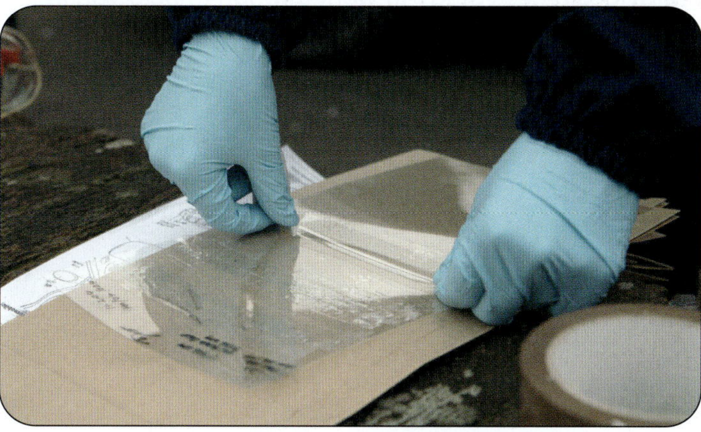

Transferring a dusted fingerprint to a lift card

The next day the police get a tip-off from a student in the local college who has heard three students, Adam (A), Belinda (B) and Charlie (C) discussing how to bypass security settings on a locked laptop.

The police go to the homes of the three students, but find no laptop. Although all three say they don't know anything about a stolen laptop, the police take away the students' trainers and ask them to come to the police station for fingerprinting.

COURSEWORK TASKS

You now need to see whether there is a link between any of the suspects and the evidence already examined. You need to:

1. Take the ***fingerprints*** of Adam, Belinda and Charlie, see page 73.
2. Make some ***impressions*** by inking the soles of their trainers, see page 78.
3. Show that your ***blood splat theory*** points to the suspect being at the table where the laptop was.

Report

1. Carefully document all your evidence.
2. Compare your evidence from the crime scene with the items recovered from the homes of the three students.
3. What conclusions can you make based on your evidence?

COURSEWORK HINTS

Pages 156–7 have the coursework checklist and mark scheme. Follow this advice. Remember this coursework counts towards 40% of your Additional Applied Science GCSE.

4.18 The case of the vindictive vandal (1)

LEARNING OBJECTIVES

1 For Unit 3 coursework, how do you investigate trace evidence using a microscope?
2 How do you test for ions in a sample solution?

For your coursework investigation in forensic science you need to carry out a number of tests and use different techniques for collecting samples of evidence. This section provides a case study to work through to help you in these tasks.

Who vandalised the car?

Setting the scene

Local farmer Terry Rout has had his garage broken into and his lovingly restored vintage car vandalised. The vandal also appears to have poisoned his trout lake. Imagine you are a forensic scientist. You have been asked to examine evidence collected by the scene of crime officer and compare this to evidence collected from each of two suspects.

Terry Rout inspecting damage to his car

The evidence

The SOCO states, "It looks like the vandal broke the lock of the garage, then smashed a hammer into the bonnet. There are no fingerprints or footprints. The hammer was clean and the floor of the building was swept with a broom. No prints on that either. I recovered some control paint fragments from the bonnet and a sample of the thick dust from the floor. I also collected some hair from Terry Rout and a sample of water from the lake."

The suspects

Police question three potential suspects that day: Mr Rout's ex-wife, Anita (A), her new partner, Barry Brown (B) and noisy neighbour Colin Clark (C). Under supervision at the police station, the following evidence is collected from each suspect:

- traces of dirt from Anita's shoes, Barry's trainers and Colin's boots
- material that the SOCO vacuums from their trousers, in turn, collecting the evidence in separate filter bags
- samples of hair from each person.

PRACTICAL

Microscopic analysis

Paint samples

Start at 100× total magnification to look at paint samples. Note the colour(s). Inspect the width of any layers. Record any features in the outer layer, such as scratches or the texture of the surface.

Dust samples

After inspection, use tweezers to break up the dust sample into hairs, fibres, soil, seeds, etc. For hairs, note their colour, shape, size, stiffness and straightness. Record your findings. Then start at a low power when viewing samples under the microscope.

Hair samples

- ***Medulla*** – This is the air-filled central channel, which varies in thickness, transparency and brokenness.
- ***Cortex*** – This is the protein-rich structure containing pigment, which varies in thickness, texture and colour.
- ***Cuticle*** – This is the outer coating of overlapping scales, which varies between animal species.

Make a ***wet mount*** of each hair by:

1 Putting a small drop of water in the centre of a microscope slide.

2 Placing a 1 cm length of hair so it lies flat on the drop of water.

3 Covering the hair and water drop with a cover slip.

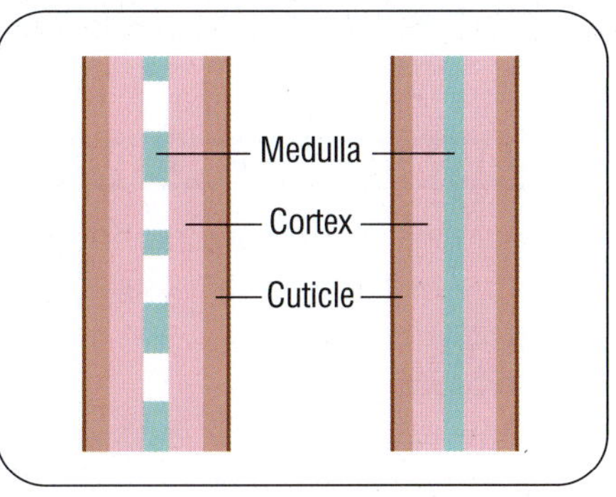

Structure of a hair follicle

Using a microscope

1 Always carry the microscope using both hands, one holding the arm and one under the base.

2 Place your slide on the stage and clip it securely.

3 Select the lowest power objective lens.

4 Before looking down the microscope wind the stage up close to but not quite touching the slide.

5 Then look down the eyepiece lens and use the focus knobs to move the stage slowly downwards until you see a clear image.

6 Repeat steps 4–5 with more powerful lenses.

- How do steps 4–5 avoid damaging the slide (and lens)?

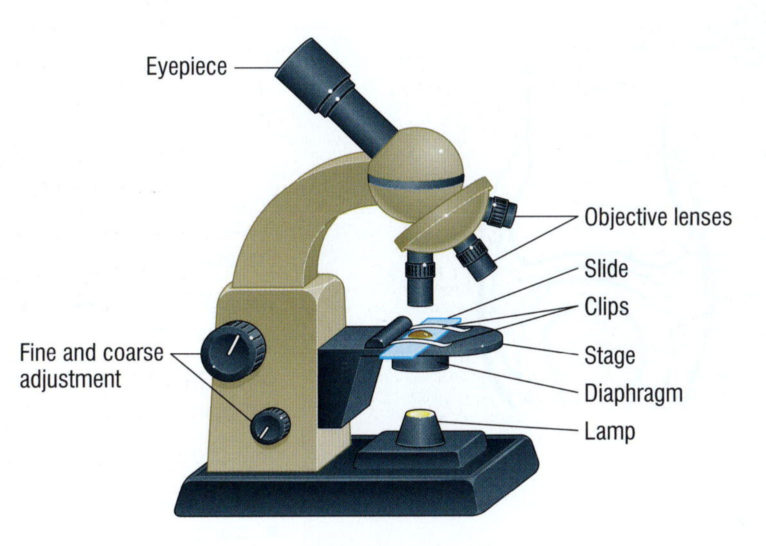

4.19 The case of the vindictive vandal (2)

LEARNING OBJECTIVES

1. For Unit 3 coursework, how do you investigate trace evidence using a microscope?
2. How do you test for ions in a sample solution?

Collecting a water sample in a tube to analyse for pollution

Who poisoned the lake?

PRACTICAL

Analysis of water sample

1. First you need to evaporate off some of the water from the sample.

   Do this in an evaporating dish over a Bunsen burner. Do not boil all the water away. The idea is to increase the concentration of any dissolved compounds in the solution.

2. Now, taking safety precautions, you need to carry out three tests:
   - The sodium hydroxide (NaOH) test for positive metal ions (see page 83 on precipitation reactions).
   - The tests for non-metal negative ions, on page 83.
   - Leave the remaining solution to evaporate away. Then carry out a flame test on any solid that remains. This may confirm the presence of the positive metal ion, found in the first test (see page 84 on flame tests).

As a result of your experiments you should be able to name any ionic compound dissolved in the water from the lake. Remember, an ionic compound contains a metal and a non-metal ion.

Safety

Take all the usual safety precautions when using Bunsen burners, acids and alkalis.

Carry out a risk assessment before you start, referring to page 17 and all the appropriate CLEAPSS Hazcards.

You are using chemicals that are:

- harmful/irritant
- corrosive

So wear eye protection.

PRACTICAL

Analysis of material samples

You need to:

1 Make up solutions from the evidence collected from the suspects (the dirt from the shoes and trainers and the materials in the vacuum filter bags from the trousers). (**Clue:** Add water, stir and filter.) Label these A, B and C.

2 See if any of these solutions contains the toxic compound. Use the sodium hydroxide test to confirm or reject a match with the positive metal ion of the toxin. Then use the appropriate test to confirm or reject a match with the negative non-metal ion.

- Do these tests indicate that any of the suspects may have polluted the lake?

COURSEWORK TASKS

1 Carry out a microscopic examination of the paint, dust and hair samples collected from the garage and from the suspects. Compare them. What conclusions can you draw from your results?

2 Analyse the water sample collected from the lake. What, if any, substances can you find that could be toxic to aquatic life?

3 Analyse the samples of material from the suspects, Anita, Barry and Colin, to see whether any of these may contain traces of any toxin found in the lake.

Report

1 Carefully document all your evidence.

2 Compare your evidence from the crime scene with the items recovered from the trousers, shoes and hair of the three suspects.

3 Compare the results from your analysis of the water sample with the results of your tests on solutions labelled A, B and C.

4 What conclusions can you make based on your evidence?

Check you have all the elements needed for a good report, shown at the top of page 3.

COURSEWORK HINTS

Pages 156–7 have the coursework checklist and mark scheme. Follow this advice to help you maximise your grades.

Sports science

Sports science can improve your performance

What you already know

Here is a quick reminder of previous work you will find useful in this chapter from AQA Science A or Science B:

- Internal body conditions that are controlled include:
  - Water content: water leaves the body via the lungs when we breathe out and via the skin when we sweat, and excess water is lost via the kidneys in urine.
  - Temperature: to maintain the temperature at which enzymes work best.
  - Blood sugar levels: to supply the cells with a constant supply of energy.
- Many processes within the body are coordinated by chemical substances called hormones. Hormones are secreted by glands and are transported to their target by the bloodstream.
- The rate at which all chemical reactions in the cells of the body are carried out (the metabolic rate) varies with the amount of activity you do and the proportion of muscle to fat in your body.
- The less exercise you take the less food you need. People who exercise regularly are usually fitter than people who take little exercise. If you exercise your metabolic rate stays high for some time after you have finished.
- Too much food and too little exercise lead to high levels of obesity.
- The hotter a body, the more thermal (infra-red) energy it radiates.
- The colour, shape and dimensions of a body affect the rate at which it transfers heat by conduction, convection, evaporation and radiation.
- The bigger the temperature difference between a body and its surroundings, the faster the rate at which heat is transferred.
- Energy is normally measured in joules (J).

RECAP QUESTIONS

1. What internal body conditions need controlling?
2. What are hormones, and how are they transported?
3. What does 'metabolic rate' mean, and what happens to it during exercise?
4. What is the name for thermal radiation, and what happens to the amount you radiate when you are hotter?
5. What are the reactants and products of aerobic respiration?
6. What is the name of the force that lets you grip, rather than slip?
7. Why do the blades of ice skates exert a large pressure on the ice?

ACTIVITY

Produce a slide for a PowerPoint presentation for a particular Olympic or Paralympic sport. Focus on one of the following:

- Motivating an athlete.
- Health, fitness or diet.
- Clothing or sports equipment.

Combine your slides into one class slideshow.

London 2012

From Key Stage 3:

- A healthy diet contains the right balance of different foods you need (carbohydrates, proteins, fats, minerals, vitamins, fibre and water).
- The role of the skeleton and joints and the movement of muscle pairs (for example, biceps and triceps).
- The role of lung structure in gas exchange.
- Aerobic respiration involves a reaction in cells between oxygen and food, in which glucose is broken down into carbon dioxide and water.
- The reactants and products of respiration are transported throughout the body in the bloodstream.
- Unbalanced forces change the speed or direction of movement of objects. Balanced forces produce no change in the movement of an object.
- Frictional forces, including air resistance, affect motion (for example, streamlining cars, friction between tyre and road).
- The relationship between force, area and pressure in, for example, the use of snowboards and skates.
- Energy is transferred by the movement of particles in conduction, convection and evaporation. Energy is transferred directly by radiation.

Making connections

Sports scientists

Sports scientists are employed to maximise the physiology of an athlete's body and the use of materials in their sport.

Energy

Energy for exercise comes from aerobic respiration (or anaerobic respiration if there is an oxygen debt).

Blood sugar levels, body temperature and water levels must be maintained if performance is not to suffer.

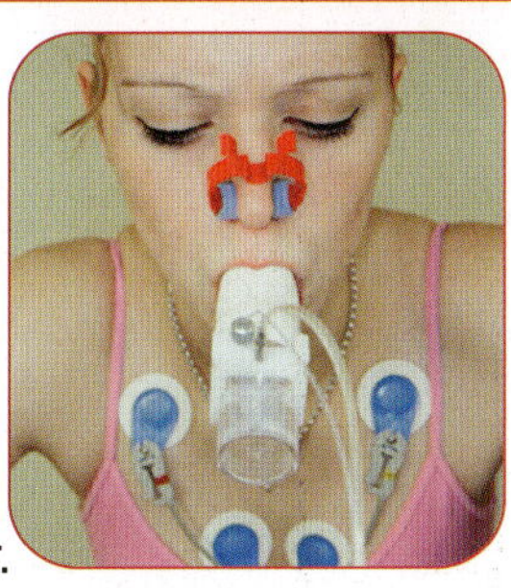

Fitness training

Fitness training increases the response of heart, lungs and muscles. Improvements can be assessed by taking base-line measurements.

Diets

Appropriate diets take into account the nature of the sport and the personal daily energy requirement of the athlete who is competing.

Clothing

Clothing and footwear have benefited from using modern materials, e.g. ***sport wool*** for football shirts is breathable, letting air in, yet moisture-absorbing, letting sweat out.

Materials

Exploiting the properties of metals, polymers, ceramics and composites has helped sports equipment designers to produce even better tennis rackets, crash helmets, etc.

5.1 Introduction to sports science

LEARNING OBJECTIVES

1. What is the role of sports physiologists, nutritionists and materials scientists?
2. What does success in sports depend upon?
3. What is antagonistic action in muscles?

The aim of the sports scientist is to improve an athlete's performance. In this chapter you will learn about some of the science and techniques used by sports physiologists, nutritionists and materials scientists:

- Health and fitness interest **sports physiologists** – in particular the parts of the body involved during exercise.
- **Nutritionists** and ***dieticians*** help to achieve peak performance in athletes, by controlling their energy and nutrient intake.
- ***Materials scientists*** study the properties of materials and develop new materials for sports clothing and sports equipment. They are interested in the effect forces have on sports equipment.

a) What is the role of a nutritionist?

Finishing a new surfboard at the Ocean Magic factory, in Newquay, Cornwall

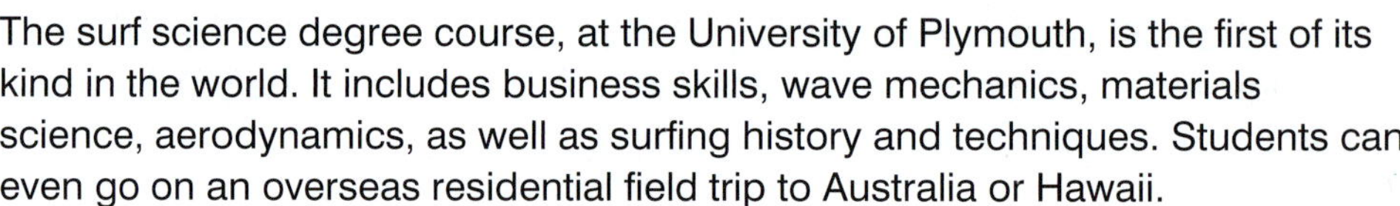

The surf science degree course, at the University of Plymouth, is the first of its kind in the world. It includes business skills, wave mechanics, materials science, aerodynamics, as well as surfing history and techniques. Students can even go on an overseas residential field trip to Australia or Hawaii.

Your success in sports depends on many factors, including:

- The fitness of your body and how it performs under stress.
- Your energy and nutrient intake before exercise.
- The effectiveness of your clothing and sports equipment.
- Your athletic skill level.
- Your ability to concentrate and focus in a competitive situation.

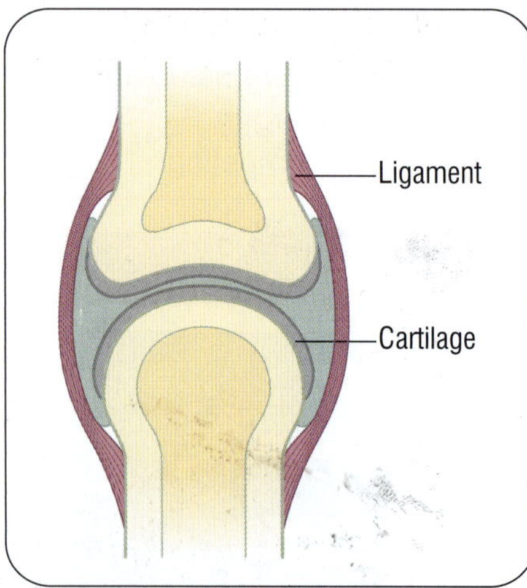

The strong fibres that hold bones together at a joint are ***ligaments***. The slippery layer that stops bones rubbing together is ***cartilage***.

Muscles and how your body moves

As humans you have three different kinds of muscles:

- ***Cardiac muscle***, causing an adult's heart to beat about 70 times every minute and pumping five litres of blood per minute.
- ***Smooth (or involuntary) muscle*** that work on their own. Examples include muscles to push food through the gut, to expel urine from the bladder and to regulate the flow of blood.
- ***Skeletal (or voluntary) muscle*** that you control as you choose. When you decide to move, skeletal muscles pull on your bones. These muscles work quickly and strongly, but they soon tire.

Using a dumb-bell

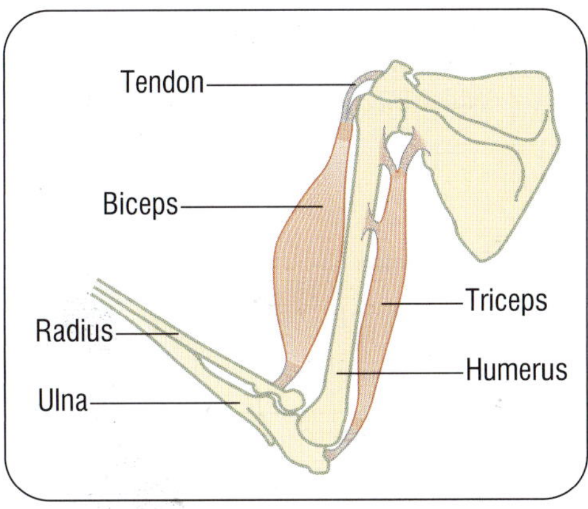

The strong fibres that attach muscles to bones are called ***tendons***. A muscle can only pull on a bone, it cannot push. For this reason muscles always work in pairs. As one muscle contracts the other relaxes. Pairs of muscles are called **antagonistic pairs**, because they always oppose each other.

A good example of an antagonistic pair is the biceps and triceps. One makes the elbow bend, the other makes it straighten.

b) Lift your forearm. Feel your biceps contract and your triceps relax. What happens to your biceps and triceps when you extend your forearm?

GET IT RIGHT!

Tendons attach muscle to bone. Ligaments attach bone to bone at joints. The elbow is a hinge joint. The shoulder is a ball and socket joint.

KEY POINTS

1 Sports physiologists look at parts of the body used during exercise.
2 Nutritionists advise on energy and nutrient intake.
3 Materials scientists study the properties of materials. They develop new materials for sports clothing and equipment.
4 Success in sport depends on fitness, equipment, skill, energy and concentration levels.
5 Skeletal (or voluntary) muscles work as antagonistic pairs.

SUMMARY QUESTIONS

1 How can you improve your

a) skill levels? b) ability to keep focused?

2 Why are at least two muscles needed at a joint?

3 What is the difference between a tendon and a ligament?

4 a) Name a hinge joint. b) Name a ball and socket joint.

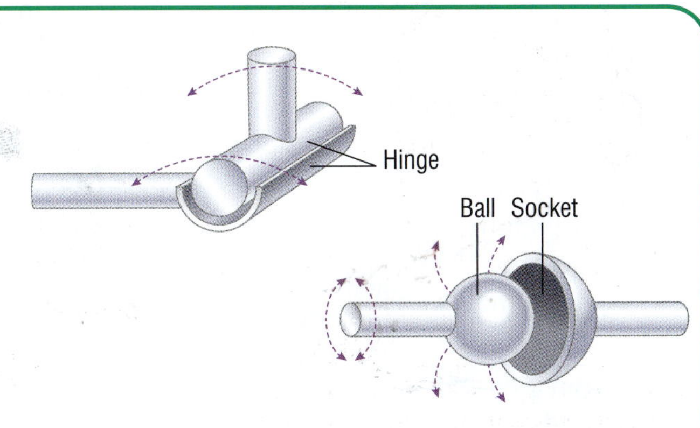

A hinge joint and a ball and socket joint

5.2 Your heart

LEARNING OBJECTIVES

1. What is the structure of the cardiovascular system?
2. How do you measure your heart rate (pulse)?

Feel your ribs. They protect your heart and lungs.

Clench your fist. Your heart has about this same size. It is a ***double pump*** made of cardiac muscle that pumps blood round your body.

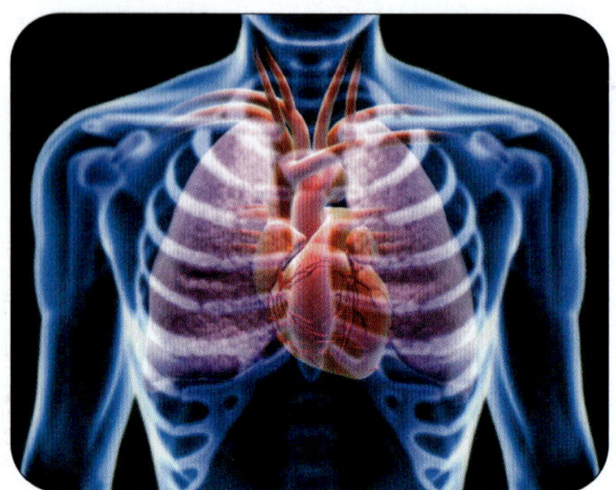
The human thorax or top part of your body

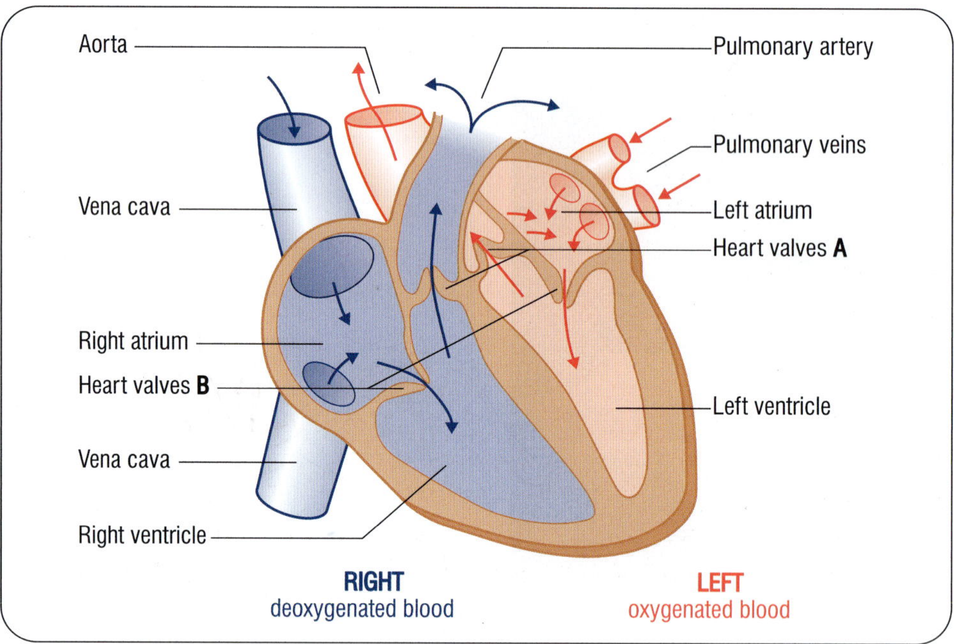

The heart has four chambers

DID YOU KNOW?

In women, the heart volume is usually about 0.5–0.6 litres. Some female swimmers have volumes of 0.9 litres.

An elite male cyclist's could be near 1.7 litres.

- The **right atrium** receives deoxygenated (O_2-poor, CO_2-rich) blood from the organs of the body.
- The ***right ventricle*** pumps this deoxygenated blood to the lungs.
- The **left atrium** receives oxygenated (O_2-rich, CO_2-poor) blood from the lungs.
- The ***left ventricle*** pumps this oxygenated blood out to all the organs (except the lungs).

a) Do all veins carry deoxygenated blood?
b) What are the lower chambers of the heart called?

When the heart beats:

- The atria gently push the blood into the ventricles. Valves B open and A close, see diagram above.
- Then the ventricles contract, forcing blood down the arteries. Valves A open and B close. The valves only allow the blood to pass through the heart in one direction.

GET IT RIGHT!

Remember:

- Blood leaves your heart through your **arteries**.
- Blood returns to your heart through your veins.
- Your pulmonary vein carries oxygenated blood ***from*** your lungs ***to*** your heart.

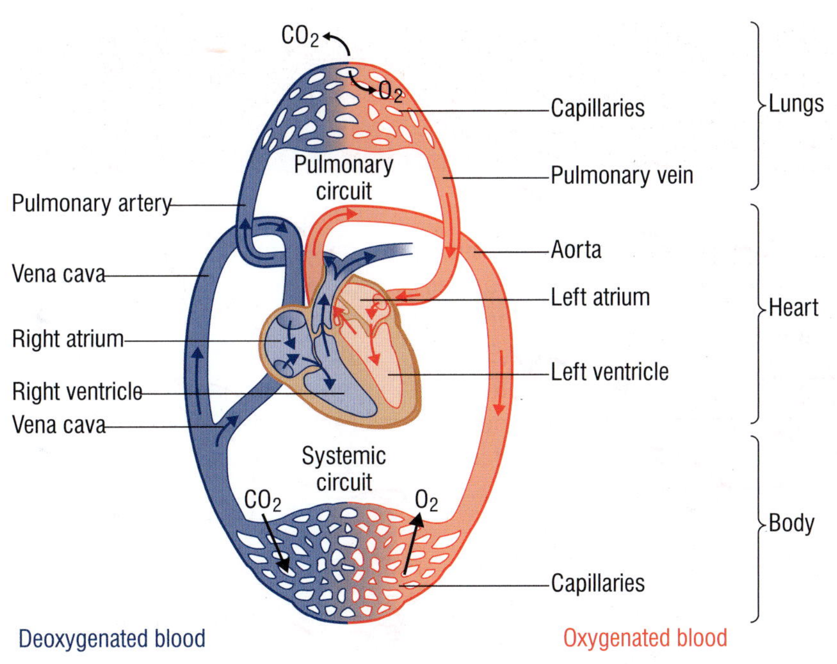

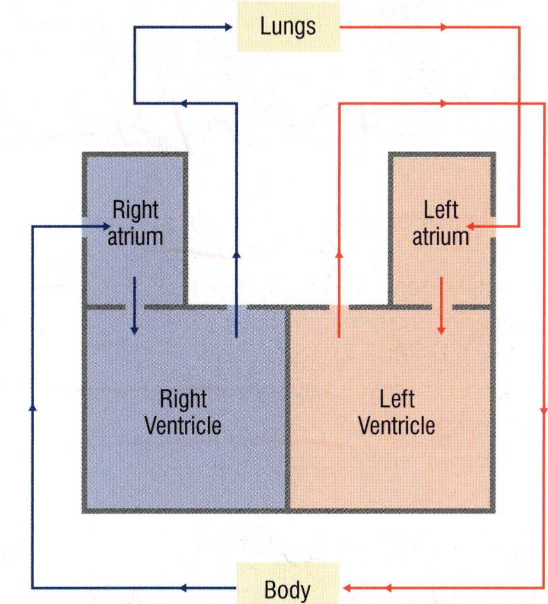

Blood circulation

PRACTICAL

Heart rate

1 After sitting quietly for five minutes, measure your heart rate. This is your resting heart rate. Either use a pulse meter or find your pulse on your neck or at your wrist (counting for 30 seconds). The average resting heart rate is 75 beats per minute.

2 Now think about one minute of step-ups or skipping and measure your pulse just before the exercise.

3 Immediately after you finish, record your heart rate again. Fill in a table:

Resting heart rate (beats per minute)	**Pre-test heart rate (beats per minute)**	**Post-test heart rate (beats per minute)**

- What may cause your pre-test heart rate to change?
- Why is your post-test rate greater than your resting heart rate?

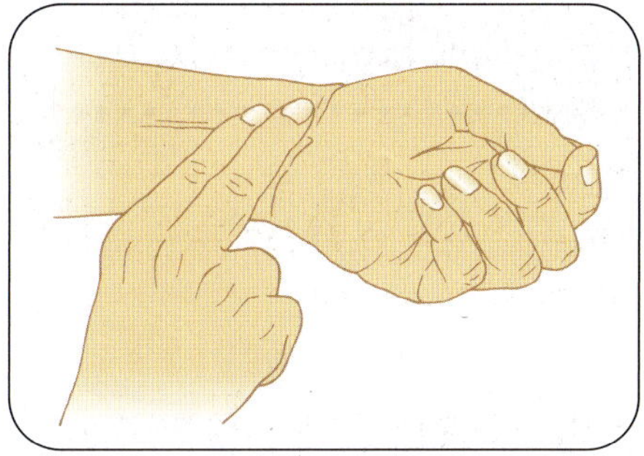

Taking your own pulse

SUMMARY QUESTIONS

1 Copy and complete: The heart is split into . . . . . . sides and has . . . . . . chambers.

2 Why is the heart called a 'double pump'?

3 a) The left ventricle has thick walls and is stronger than the right ventricle. Why is this?
 b) Why are ventricles thicker-walled than atria (the plural of atrium)? (Think where the blood goes.)

4 Some people have larger hearts. How does this help in sports?

KEY POINTS

1 Your cardiovascular system is your heart and its blood vessels.

2 Blood leaves your heart's ventricles through arteries. Blood returns to your heart's atria through veins.

3 Your right ventricle pumps deoxygenated blood to your lungs. Your left ventricle pumps oxygenated blood to the rest of your body.

5.3 Your lungs

LEARNING OBJECTIVES

1 How does the structure of the thorax enable ventilation of the lungs?
2 How do you measure your breathing rate?
3 What is vital capacity and tidal volume?

DID YOU KNOW?

You have about 300 million alveoli, with a **surface area** equivalent to half a tennis court!

GET IT RIGHT!

Remember when you breathe in your lungs increase in size because your:

- Diaphragm contracts and moves down.
- Intercostal muscles contract and lift your ribs up and out.

Your lungs are like big sponges that exchange the gases oxygen and carbon dioxide. At the ends of the bronchioles are air sacs, made up of 'grapes' called **alveoli**. Round the thin-walled alveoli are blood capillaries. Here oxygen from the air passes into the red blood cells. Similarly carbon dioxide leaves. The **deoxygenated blood**, that arrived at the lungs in the pulmonary arteries (blue in the diagram on the previous page), now returns to the heart as ***oxygenated blood*** through the pulmonary vein (red in the diagram).

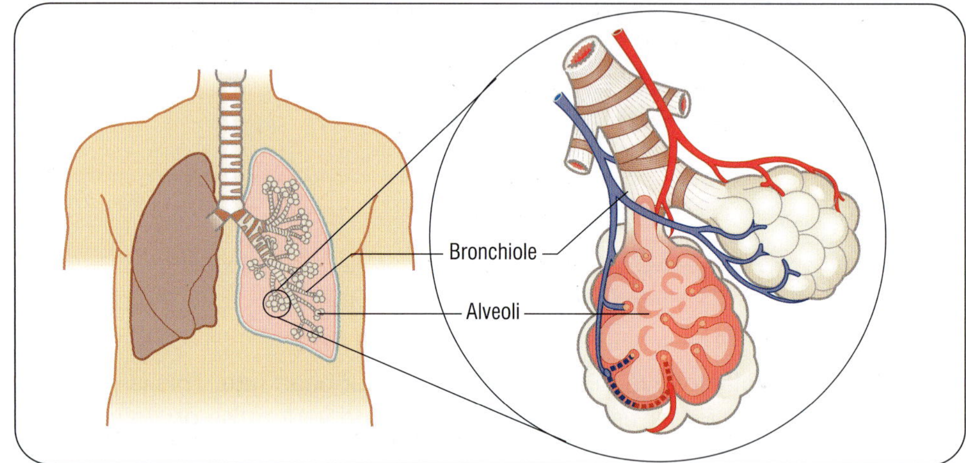

The lungs and air sacs

Breathe in and out. Feel your sternum (breastbone) and your diaphragm. As the muscles in your **thorax** relax, your lungs decrease in size, and you breathe out. However when breathing in, the volume of your lungs increases because:

- the **diaphragm** contracts and moves down,
- the ***intercostal muscles***, between your ribs, contract and pull your ribcage and sternum up and out.

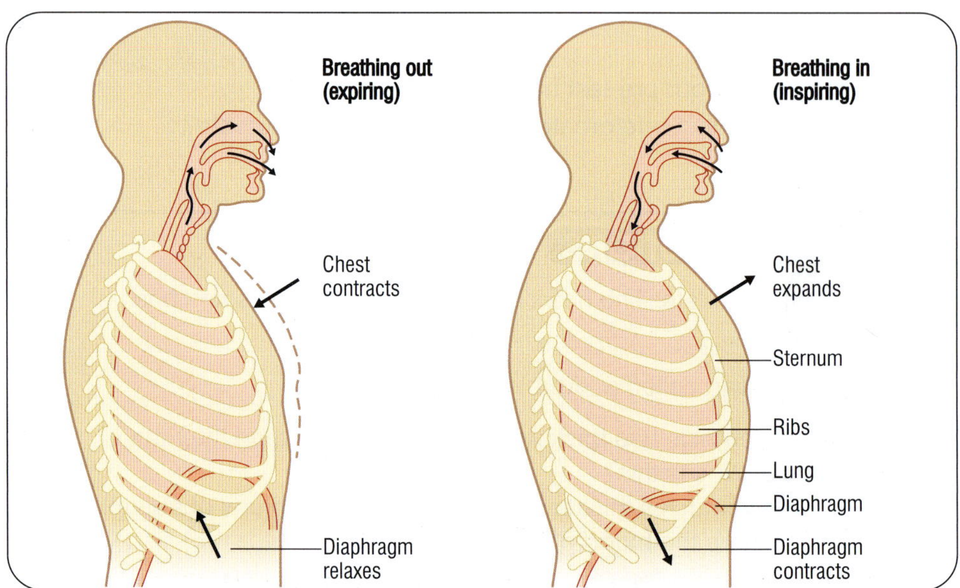

Breathing out and breathing in

a) Where in the lungs does gas exchange take place?
b) What does your diaphragm do to inflate your lungs?
c) Where are the intercostal muscles?
d) How do your intercostal muscles aid ventilation of your lungs?
e) Breathing **ventilates** your lungs with fresh air containing 21% oxygen. You exhale air containing only 16% oxygen. How much extra carbon dioxide is breathed out?

Breathe out
Relax
Diaphragm relaxes
Breathe in
Intercostal muscles contract and ribs rise
Diaphragm contracts

Breathing

PRACTICAL

Breathing – Measuring lung volumes with a digital spirometer

Taking appropriate precautions and wearing suitable clothing, take the following measurements before and after exercise, such as a 400 m race:

1 ***Breathing rate*** in breaths per minute (counting for one minute). The average resting breathing rate is 12 breaths per minute.

2 ***Tidal volume (TV)***. TV is the volume of air you breathe in and out normally. Measure this volume in ml (or cm^3) using a spirometer.

3 ***Vital capacity (VC)***. VC is the maximum volume of air you can force out after breathing in as hard as you can, or the volume you can expire after maximum inspiration.

There is not much difference between the lung capacities of athletes and non-athletes. However the amount of air that an athlete can breathe in and out in one breath (the tidal volume) does increase slightly with exercise. This is due to the intercostal muscles becoming stronger.

SUMMARY QUESTIONS

1 What is inspiration?

2 What two muscles help you to breathe in?

3 How does the large surface area of your lungs help when breathing?

4 How do your lungs and heart work together to get oxygen to your muscles?

5 How does a greater tidal volume help an athlete?

6 Why is more carbon dioxide breathed out than breathed in?

KEY POINTS

1 Your diaphragm and the intercostal muscles between your ribs help you to ventilate your lungs or breathe.
2 If these muscles in your thorax relax, your lungs decrease in size and you breathe out.
3 When you breathe in your lungs increase in size because:
- Your diaphragm contracts and moves down.
- Your intercostals contract and lift your ribs up and out.

5.4 Changes during exercise

LEARNING OBJECTIVES

1 How do the heart and lungs help to provide glucose and oxygen to the muscles?
2 What are aerobic and anaerobic respiration?
3 What changes occur to breathing and heart rate during exercise?

All athletes need a diet that provides enough energy. **Carbohydrates** give us energy. We get carbohydrates from **sugars** and **starch**. Bread, pasta, rice and potatoes contain starch.

As you digest carbohydrates, they break down into small glucose molecules. Your bloodstream **absorbs** glucose from your gut and carries it to your muscles. Glucose is the only carbohydrate that muscles can use directly for energy.

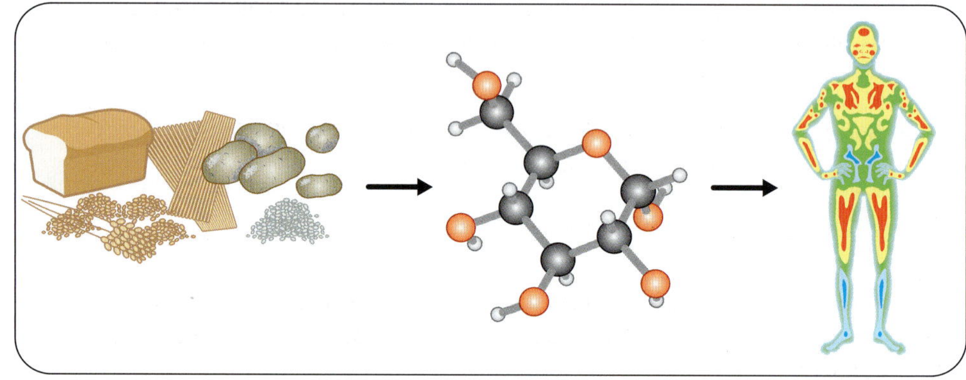

Starch to glucose ($C_6H_{12}O_6$) to energy

GET IT RIGHT!

Respiration is ***not*** breathing in and out. Respiration is the ***process*** of releasing energy from glucose.

Respiration is the process of releasing energy from glucose. Respiration goes on in every cell of your body. Muscles need energy to contract.

Aerobic respiration

Aerobic respiration needs oxygen. 'Aerobic' means 'with oxygen'. This happens when you are jogging. Jogging does not need much energy, and there is plenty of time for your red blood cells to carry all the oxygen you need from your lungs to your muscles.

a) What kind of respiration needs oxygen?

You can write aerobic respiration as:

glucose + oxygen → carbon dioxide + water (+ energy)

$$C_6H_{12}O_6 + 6O_2 \rightarrow 6CO_2 + 6H_2O \ (+\ \text{energy})$$

The aerobic respiration of 1 g of glucose produces 16.1 kJ of energy.

Carbon dioxide and water are waste products that we must get rid of. We store water in our bladders before we excrete it. We get rid of carbon dioxide from our lungs.

With training, endurance athletes develop more (and wider) blood capillaries around their alveoli. Their blood can now flow past their lungs more freely, and they can exchange gases at a faster rate.

b) What gases do you exchange in your lungs?

Anaerobic respiration

Anaerobic respiration does not need oxygen (see also page 22).

This happens when you are running fast, as your muscles need more energy than they can get from aerobic respiration. No matter how fast you breathe, or how fast your heart beats, you cannot get oxygen to your muscles quick enough. You have an **oxygen debt** and anaerobic respiration takes over:

glucose → lactic acid **(+ energy)** (Look, no oxygen!)

Anaerobic respiration only produces 1.2 kJ for every gram of glucose.

Lactic acid ($C_3H_6O_3$) is not good either. It makes your muscles ache.

In sport, both aerobic and anaerobic respiration work together to provide energy. However as your exertion increases, aerobic respiration decreases. In a long distance race you may use 70% aerobic energy. In a sprint aerobic respiration is almost zero.

However, aerobic training increases your ability to get more oxygen to your muscles, which helps in endurance events.

c) Why does aerobic training help you to exercise for longer without tiring?

If a person stops training then, over a period of time, their body reverts to its pre-training state. There will be a loss in muscle mass and a decrease in oxygen delivery.

d) What happens to athletes who are injured for some time?

SUMMARY QUESTIONS

1 Copy and complete:

S . . . . . . and s. . . . . . are carbohydrates. These break down to g. . . . . . when digested, which respiration converts to e. . . . . ..

2 If more oxygen enters your bloodstream each second, what happens to the oxygen reaching your muscles?

3 What is the difference between aerobic and anaerobic respiration?

4 From what sort of respiration does a 400 m runner get most of their energy?

5 a) You know that more exercise makes your muscles larger. What happens to the size of your heart as you exercise?
  b) A larger heart pumps out more blood every beat. How does this affect the oxygen delivered to your muscles every second?
  c) How does aerobic training increase your ability to get oxygen to your muscles?

DID YOU KNOW?

A normal resting heart rate is about 70 beats per minute. However rates as low as 50 beats per minute are normal in athletes. Exercise has improved their hearts' ability to pump blood. So now fewer heart beats supply all the oxygen that they require.

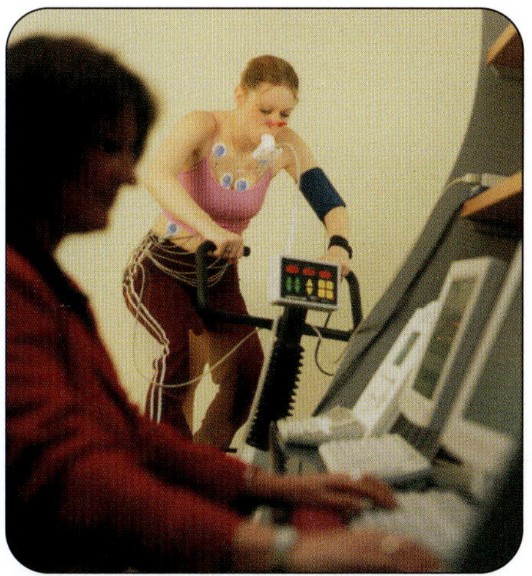

Physiological stamina test. The athlete is breathing through a mouthpiece to assess her oxygen consumption. Electrodes monitor her heart activity and blood pressure.

KEY POINTS

1 The heart and lungs provide glucose and oxygen to the muscles.
2 Aerobic respiration needs oxygen. Anaerobic respiration does not need oxygen.
3 Anaerobic respiration takes over when an oxygen debt occurs.

5.5 Recovery after exercise

LEARNING OBJECTIVES

1. What is oxygen debt?
2. How do you control your blood glucose levels?

On the previous page, you learnt that anaerobic respiration releases energy from glucose without the use of oxygen. This happens in your muscles during strenuous sports, when oxygen cannot be supplied quickly enough for aerobic respiration. During anaerobic respiration, the build up of lactic acid makes your muscles ache.

DID YOU KNOW?

Anaerobic respiration (glucose → lactic acid (+ energy)) only gives about $\frac{1}{13}$th as much energy as aerobic respiration.

Oxygen debt

After vigorous exercise you have an 'oxygen debt' because your lungs and heart rate could not keep up with the demand earlier on. You need oxygen to break down lactic acid into harmless carbon dioxide and water.

Before your muscles can operate effectively again, you must:

1 Remove lactic acid.
2 Replace energy.
} takes about 20 minutes

3 Top up haemoglobin with oxygen.
4 Replenish stores of glycogen. (More on this later.)
} takes over 24 hours

Warming down or jogging slowly after you exercise keeps your heart rate and breathing rate up. An active recovery ***repays the oxygen debt***, removes lactic acid and replaces your energy more quickly. More lactic acid remains in your blood if you don't warm down actively.

a) Why can you feel sore up to two days after intense exercise?

Glycogen

Your body cannot store glucose in your blood. Instead you store it as a starch, called **glycogen**, in your liver (20%) and your muscles (80%).

The liver is the largest **organ** in your body. Two hormones, **insulin** and **glucagon**, control the amount of glucose in your blood. These hormones (or chemical messengers) are made in your pancreas.

Diabetes is a disease where people don't produce enough insulin. Their blood sugar levels can rise to dangerous levels, causing blackouts and blindness.

Liver
Pancreas

The liver and pancreas

This girl is injecting herself with insulin. This makes her liver remove the glucose from her blood. Too much insulin could make her dizzy. A sweet can give her a quick glucose boost.

Controlling blood sugar levels

Your pancreas monitors and controls the glucose concentration in your blood:

- Digesting carbohydrate foods puts glucose into your blood. When there is too much glucose in your blood, your pancreas secretes the hormone insulin. Insulin makes your liver convert soluble glucose into insoluble glycogen and store it.
- Exercise removes glucose from your blood. When there is too little glucose in your blood, your pancreas secretes the hormone glucagon. Glucagon makes your liver convert glycogen into glucose, and release it into the blood stream.

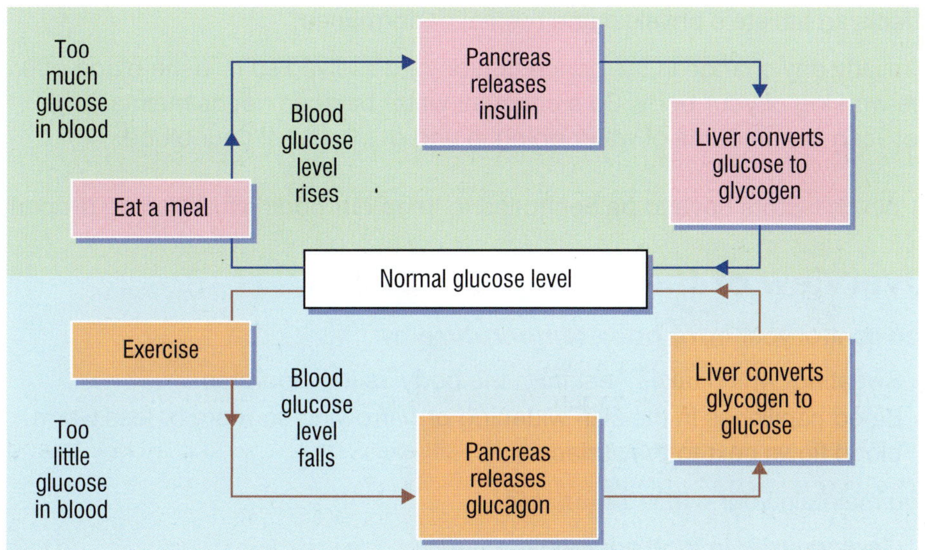

The control of blood sugar levels

b) i) Which two hormones control the concentration of blood glucose?
ii) Where are these hormones made?

Athletes taking insulin, like Olympic gold-medal rower, Sir Steve Redgrave, take special precautions before starting a workout programme. That's because they need to control their blood glucose levels carefully.

c) What happens to the concentration of blood glucose when not enough insulin is produced? Why is this dangerous?

SUMMARY QUESTIONS

1 Which kind of respiration occurs in your body when:
a) You are sitting still? b) You are running fast?

2 What substance makes your muscles ache after hard exercise?

3 What do we mean by an 'oxygen debt'?

4 Why does your breathing rate and pulse remain high while you recover after exercise?

5 What does insulin do when your blood sugar levels are too high?

6 People who are inactive, overweight and older, or who have a family history of diabetes, have an increased risk of developing diabetes. What role has exercise in preventing diabetes?

KEY POINTS

1 Anaerobic respiration takes over when an oxygen debt occurs.
2 Your body cannot store glucose in your blood. Instead you store it as a starch, called glycogen, in your liver (20%) and your muscles (80%).
3 The hormones insulin and glucagon control your blood glucose levels.

5.6 Controlling temperature and water levels

LEARNING OBJECTIVES

1 How do you maintain a constant body temperature?
2 How do you maintain the correct amount of water in your body?

Your body works best when the conditions inside it (the internal environment) remain the same.

For sports scientists the ear is the most common location to measure core temperature. The core body temperature in humans is 37 °C, although the skin temperature varies. An increase or decrease in the core temperature by 1 °C affects an athlete's physical and mental performance.

Similarly any change in the concentration of dissolved salts in the blood affects the working of your body. So a constant water balance needs maintaining, between the amounts of water going in and going out of your blood.

a) Which factors have to be controlled to keep our bodies functioning properly?

Overview of the main processes

You control your ***core body temperature*** by:

- Sweating, which takes heat from the body as it evaporates.
- Blood capillaries in the skin widening or narrowing so more or less warm blood flows next to the surface.

You maintain your ***water levels*** by:

- Gaining water in food and drink.
- Losing water in urine and sweat.

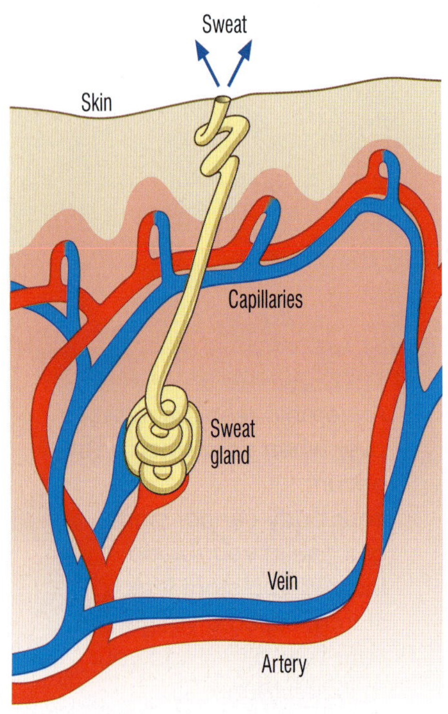

Your skin helps to control your temperature

PRACTICAL

If your core body temperature is ***too hot***:

When you get out of the shower you feel the effect of **evaporation** causing you to cool down. When you get too hot you sweat a lot! The sweat also takes energy from your skin as the water evaporates.

Feel the saliva in your mouth with your tongue.

- Is it cold?
- Now lick the back of your hand. Does it feel warm now?
- Where does the water get energy from to evaporate?

When you get too hot the blood vessels supplying your **capillaries** just below your skin expand. More blood flows near the surface. Then the heat in your blood **conducts** to the surface where it **radiates** away.

Your face is probably warmer than the table you are sitting at. Hold one hand near your face (but not touching), the other near the table. Which hand detects the most infra-red radiation?

If your core body temperature is ***too cold***:

When you get too cold you stop sweating. Think of the colour of your skin when you are cold. When you get too cold the blood capillaries just below your skin contract. Now very little oxygenated red blood flows near the surface, which limits the heat loss by radiation.

b) What two changes in your skin help to keep you cool during exercise?

Think of a confused weak mountaineer on an exposed rock-face. Hypothermia has set in ('hypo' means low). Shivering can no longer maintain his body temperature. His skin alone won't keep him warm in the cold. He hasn't got thick insulating hair that traps air like this polar bear!

Polar bears have thick insulating hair

Competing in a hot humid climate

Heat loss by radiation is not possible if the environment is hotter than the athlete. Similarly, the more humid the environment the less the athlete's sweat evaporates.

If an athlete cannot cool down by radiation or evaporation, their core body temperature will increase by 1 °C every six minutes. Muscle cramp and heat exhaustion are likely.

Fortunately pouring cold water over you helps. In fact this is 25 times more effective in cooling you down than your body is capable of naturally. A good breeze also helps, as it aids the evaporation of water in sweat. Blow across your hand and see. That's why we use fans!

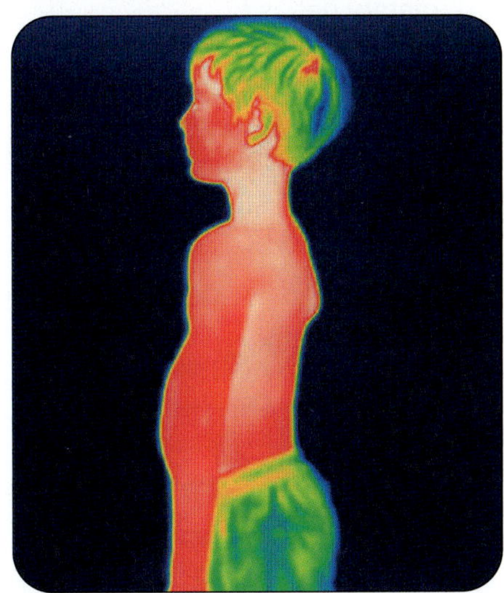

Thermogram of boy emitting radiation after playing sports

Controlling the amount of water in your blood

You control your own water input. Your sweat glands and kidneys control your water output for you. Kidneys are filters that ***clean*** the blood as it flow past. Your kidneys dump toxins and excess water into the urine. You store urine in your **bladder** until you empty it in the toilet.

c) Where is urine made?
d) What is urine mainly made of?

SUMMARY QUESTIONS

1 Why do your core body temperature and blood water levels need to remain steady?

2 Muscles put water into your blood during aerobic respiration. In what two ways can this water leave your body?

3 Why does sweating keep you cool?

4 How do insulating clothes help to keep skiers warm?

5 Imagine it is a hot day. You sweat a lot playing a sport. Why is your urine a dark concentrated colour when you go to the toilet?

KEY POINTS

1 During exercise your core body temperature increases. You sweat and blood capillaries below your skin expand. Heat loss by evaporation and by radiation cause cooling.

2 You remove water in your body in sweat and urine.

5.7 Sports nutrition

LEARNING OBJECTIVES

1 How do people's energy requirements depend on their weight and their level of exercise?
2 How do athletes record their dietary habits?

High protein breakfasts of a powerful sprint athlete and a weightlifter

What athletes eat on a day-to-day basis is important for their training. Only with a correct combination of carbohydrates, proteins, fats, vitamins, minerals and water can an athlete reach their full potential.

You have probably seen on page 32 that your body requires a variety of nutrients in order to carry out the vital functions of life: respiration, movement, growth and repair of body tissue.

- **Carbohydrates**, from sugar and starch, provide energy. Every physical activity requires energy. Hard exercise uses up to 4 g of glucose every minute. Glycogen (see page 124) is the main source of fuel used by muscles. If you train with low glycogen stores you will feel tired, your performance will be lower and you will be more prone to injury and illness.

a) Why do more strenuous activities burn up more energy?
b) Why is it necessary for athletes to take in enough energy every day?

- **Proteins** aid the growth and repair of body tissues, and provide energy. You get protein in meat, fish, eggs, milk, cheese, cereal, peas and beans. Recommended protein requirements for people aged 15 and over is 0.8 g per kg of body weight every day.

- **Saturated and unsaturated fats** supply you with energy and insulation. No more than 30% of the energy should come from fat and no more than 15% from protein.

c) Look at the breakfasts – a fried egg sandwich and eggs, burgers and cheese. Why could they be unhealthy breakfasts? See page 31.

- **Vitamins and minerals** protect vital organs by keeping them working efficiently.

- **Fibre**, from plant cell walls, helps to move waste through your intestine. Waste materials contain ***toxins***, i.e. poisons.

- ***Water*** helps to remove toxins from the blood. Insufficient water in athletes causes heat exhaustion, signaled by cramp, nausea and headache. Squash, fizzy drinks and fruit juices provide water, but one-third of your water intake comes from fruit and vegetables. If your urine is dark yellow, you need more fluid in your diet.

Dietary intake

Sports nutritionists and dieticians study the nutrient intake of athletes. Their advice can help an athlete to maximise his or her performance. The simplest way to monitor your diet is to keep a diary.

You can use a diary to track how many kilocalories (kcal) or kilojoules (kJ) of energy you eat from each source of food. (1 kcal = 4.2 kJ.)

For some foods you will need to look up the information in a food table. (For an example on the Internet put '*sports coach food composition tables*' into a Google search.) Food tables also show details of the energy (in kcal), protein (in g), carbohydrate (g) and fat (g) for 100 g of the food.

However, today most foods have kcal and kJ written on the nutrition label of their wrappers.

d) i) Look at the nutrition label. How much energy do two sheets of lasagne provide?
ii) Look at the 'energy value' line and the units of kcal. What is the average daily energy need for a woman? Include the units.
iii) A sheet of lasagne has a mass of 16 g. What percentage of protein is in a sheet of lasagne?

NUTRITION

TYPICAL VALUES	PER SHEET (APPROX 16 g)	PER 100 g (UNCOOKED)
Energy Value (calories)	230 kJ (55 kcal)	1470 kJ (345 kcal)
Protein	2 g	12 g MEDIUM
Carbohydrate (of which sugars	12 g 0.3 g	72 g HIGH 2 g) LOW
Fat (of which saturates	0.2 g Trace g	1 g LOW 0.3 g) LOW
Fibre	0.5 g	3 g MEDIUM
Sodium	Trace g	Trace g LOW

GUIDELINE DAILY AMOUNTS

	Women	Men	Approx. Per sheet
Calories	2000	2500	55
Fat	70 g	95 g	0.2 g
Salt	5 g	7 g	Nil

These figures are for average adults of normal weight. Your own requirements will vary with age, size and activity level.

Nutrition label on a packet of lasagne

SUMMARY QUESTIONS

1 Copy and select.
People need ***more*** / ***less*** carbohydrate for energy if they weigh more and do more exercise.

2 The energy content of ice cream is 170 kcal per 100 g. Suppose you only eat 1 g (unlikely), its energy is 1.7 kcal. 25 g is more likely. How much energy will 25 g of ice cream give you (in kcal and kJ)?

3 a) Why shouldn't you leave more than 24 hours between eating and recording the diary information?
b) Copy the table and record everything you eat and drink in the next 24 hours.

Food	**Quantity**	**Total energy (kcal)**	**Carbohydrate (g)**	**Fat (g)**	**Protein (g)**

Weigh out quantities of food like breakfast cereal first. Note the information on the nutrition labels. Recall any foods you didn't have time to note down at the time. Then use a food table to find the data you need.

c) i) Weigh yourself in kg. Multiply your mass in kg by 0.8. This is your ***target*** protein intake in grams.
ii) Total your protein column in grams. Have you eaten more or less protein than your 0.8 target?
d) Total your carbohydrate and fat columns. What percentage of these two is fat? Is it more than 30%? Is this good or bad? Explain why?

KEY POINTS

1 Only with a correct combination of nutrients can an athlete reach their full potential.
2 Your daily energy requirement depends on your mass and level of exercise.
3 The simplest way to monitor your diet is to keep a diary.

5.8 Energy requirements

LEARNING OBJECTIVES

1 How do people's energy requirements depend on their weight and their level of exercise?
2 Why do athletes increase their intake of complex carbohydrates before competing?

Every physical activity requires energy. The amount depends on the duration and type of your activity, as well as the mass of the person.

What is your personal daily energy requirement?

Your **basic energy requirement** (for every kg of body mass) or

$$\begin{aligned}\textbf{BER} &= 1.3\,\text{kcal per hour}\\ &= 1.3 \times 24\ \text{hours}\\ &= 31.2\,\text{kcal every day}\end{aligned}$$

a) What is the BER for a 65 kg man in 24 hours? Show that this is about 8500 kJ/day.

GET IT RIGHT!

BER = kg x 1.3 kcal x hours

1 kcal = 4.2 kJ

The extra energy you require when exercising depends on the type of activity and the energy that you expend.

To run 1 km you need an extra amount of energy = (your mass in kg) × 330 kJ.

We know the energy cost of many sports activities like this by monitoring the oxygen consumption of athletes. For example to run 1 km you must breathe in an extra 15.6 litres of oxygen (per kg approximately).

Your body needs ***more*** energy (and more carbohydrate) than a basic amount if you train or work hard physically. Your basic need is called your basic energy requirement (BER). So:

Your personal energy requirement = BER + an extra amount

For every kg of body weight you need a BER of 31.2 kcal every day.

For every hour you exercise you need an ***extra*** 8.5 kcal for each kg of body weight.

Suppose you weigh 50 kg and exercise for two hours today:

$$\begin{aligned}\textbf{Your basic need BER} &= 50 \times 31.2\\ &= 1560\,\text{kcal}\end{aligned}$$

$$\begin{aligned}\textbf{Your extra amount} &= 2 \times 8.5 \times 50\\ &= 850\,\text{kcal (i.e. hours} \times 8.5 \times \text{kg)}\end{aligned}$$

$$\begin{aligned}\textbf{Then your personal energy requirement} &= 1560 + 850\\ &= 2410\,\text{kcal}\end{aligned}$$

b) Why does an athlete require a higher carbohydrate diet than the average person?

Tae kwon do athletes breathing into masks to measure their oxygen consumption

DID YOU KNOW?

For every extra 7 kg of body mass, you expend 10% extra energy exercising.

PRACTICAL

Weigh yourself. What is your body mass in kilograms?

- What is your ***basic*** daily energy need?
- How much exercise did you do yesterday? (Answer in hours.)
- What ***extra*** energy did you need?
- What was your ***personal energy requirement*** yesterday?

Energy fuel: carbohydrates, fats, proteins

Energy fuel

Suppose you need 2410 kcal today. The ideal energy fuel blend for your body is:

Ideal fuel blend	Energy per gram	Energy share (kcal)	Mass of each (g)
57% carbohydrate	4 kcal	$0.57 \times 2410 = 1374$	$1374 \div 4 = 343$
30% fat	9 kcal	$0.30 \times 2410 = 723$	$723 \div 9 = 80$
13% protein	4 kcal	$0.13 \times 2410 = 313$	$313 \div 4 = 78$
		Total = 2410 kcal	Total = 501 g

Nutritional tips

1. Don't skip meals. It doesn't help to control your weight.
2. Most of us don't drink enough water. Aim to have six to eight glasses a day.
3. 'Eat up your greens!' Fruit and vegetables keep you healthy. Aim for five pieces each day.
4. 'Live bio' yoghurt puts good bacteria in your gut to aid digestion.
5. Buy 'low fat' dairy products to avoid the fat.
6. 'A *little* of what you fancy does you good', but alcohol is not a fuel for exercise.

SUMMARY QUESTIONS

1. In which sports events do athletes need large amounts of carbohydrate in their diet?
2. Protein, when combined with exercise, helps to grow muscle. Which types of sportsmen and women need a high-protein diet?
3. Why do athletes increase carbohydrate intake and taper off training before competition?
4. Imagine it is the morning of a competition. You do not feel like eating.
   a) Why might you feel like this?
   b) Why should you eat anyway?
5. Complete an energy fuel table for your personal energy requirement yesterday.

KEY POINTS

1. All athletes need a ***carbohydrate-rich*** diet that provides enough energy.
2. Your daily energy requirement depends on your mass and level of exercise.
3. Carbohydrates break down into glucose, which your muscles use for energy.

5.9 Sports drinks and sports diets

LEARNING OBJECTIVES

1 What do isotonic sports drinks contain?
2 How does a normally balanced diet compare with the diets of different athletes?
3 Why do athletes increase their intake of complex carbohydrates before competing?
4 Why do some athletes eat a high protein diet?

Imagine you are dehydrated after sweating. Water is not the best drink to replace your lost body fluid. It quenches your thirst before you have drunk enough. Water also encourages your kidneys to produce urine, which delays re-hydration.

Your body contains 50–75% water depending on your age and body fat. The average man needs 2.9 litres of water each day; the average woman needs 2.2 litres. The amount of water you need varies with the climate and your level of physical activity.

Nowadays sports drinks are a big business. They are specially formulated carbohydrate drinks made from water, glucose, electrolytes (containing ions) and flavourings. They re-hydrate and also boost glucose energy to working muscles. This maximises performance and endurance.

Five factors are important in designing a sports drink:

- ***Taste*** – Fruit flavours make the drink appetising.
- ***Glucose content*** – This provides instant fuel for energy. Jockeys and gymnasts who sweat, but want to maintain a low body weight, prefer less carbohydrate in their drink.
- ***Stomach emptying*** – The more you drink the faster the drink enters your intestine. Drinks with more than 7 g of glucose in every 100 g of water, empty slower than water alone.
- ***Fluid absorption*** – Glucose and ions stimulate absorption of fluid into the blood stream.
- ***Urine production*** – Sodium and potassium salts in the drink reduce urine output. Any solution containing dissolved ions like Na^+ and K^+ is called an ***electrolyte***. Electrolytes conduct electricity. Sports drinks are often called '**isotonic**' or 'isoelectronic'. *Iso* means *same*. The same proportion of ions in the drink and in the blood is the ideal.

DID YOU KNOW?

A 2% loss of body fluid saps performance. A 5% loss causes heat exhaustion.

Isotonic sports drinks

PRACTICAL

Make your own isotonic drink.

Safety: Ensure your equipment is thoroughly sterilised.

a) What are the three main ingredients of isotonic sports drinks?
b) Why should your drink contain less than 8% glucose?
c) For flavouring, why should you use reduced-sugar squash?

All athletes need a carbohydrate-rich diet that provides enough energy.

Carbohydrate / glycogen loading

Carbohydrates break down into glucose, which your muscles use for energy. You store glucose as glycogen in your liver and muscles.

Your body only stores enough glycogen for 90 minutes of exercise.

Athletes who 'hit the wall' have depleted their glycogen stores. They must consume carbohydrates during a marathon to maintain their blood glucose levels.

A sports drink replaces both the carbohydrate used for respiration and the water lost as sweat. However a 'carbohydrate/glycogen loading' diet boosts the athlete's levels of glycogen and super-compensates for any loss during their race.

Before event	Diet	Exercise
Days 7–5	Low-carbohydrate diet	Strenuous exercise
Day 4–2	Increase carbohydrates	Reduce exercise
Day 1	High-carbohydrate diet and 'pasta party' the night before	Complete rest day

d) Why does this diet allow athletes to exercise for longer?

Paula Radcliffe shattering her previous world-best in the 2003 London Marathon

High-protein diets

The longer and the more intensely an athlete trains, the more their protein breaks down. Unless carbohydrate intake is sufficient to meet their energy needs, protein will get used for energy rather than for growth and repair.

However, eating more protein will not increase the size of your muscles! Muscles develop from exercise when there is enough protein in your diet. Power athletes, such as sprinters, train to gain muscle mass. Their diet includes:

Breakfast: Bacon and eggs.

Lunch: Salad and double cheeseburger (without bread).

Dinner: Steak (or fried chicken or fish), with a salad topped in cheese dressing.

A high-protein low-carbohydrate diet puts stress on the kidneys, and can cause headaches, tiredness, dizziness and constipation.

Sprint athlete

SUMMARY QUESTIONS

1 Why do athletes sweat during exercise?

2 Why do marathon runners need isotonic drinks during a race?

3 A banana snack, with plenty of water, one hour before a competition is good. Why?

4 Why does a high-protein diet encourage the over-eating of saturated fat and cholesterol, increasing the risk of heart disease?

KEY POINTS

1 Isotonic sports drinks contain water, glucose and electrolytes.

2 All athletes need a carbohydrate-rich diet that provides enough energy.

3 Power athletes, such as sprinters, train to gain muscle mass. They have a high-protein diet.

5.10 Base-line measurements

LEARNING OBJECTIVES

1 What are base-line measurements?
2 What does body mass index indicate?
3 How do you measure muscle strength, reaction time, and the glucose content of blood or urine?

Before sports scientists advise an athlete on an appropriate fitness programme, they take **base-line measurements** of changes before, during and after exercise. Measurements are taken such as pulse (see page 117), breathing rate, tidal volume and vital capacity (see page 119), weight, body mass index (BMI), muscle strength, reaction time, and the glucose content of blood or urine.

Body mass index (BMI)

Being overweight increases your risk of heart disease and other illnesses. Body mass index is a good indicator of total body fat. **Body mass index (BMI)** reveals your ideal weight. Your BMI takes into account both your weight (or mass m in kg) and your height (h in m):

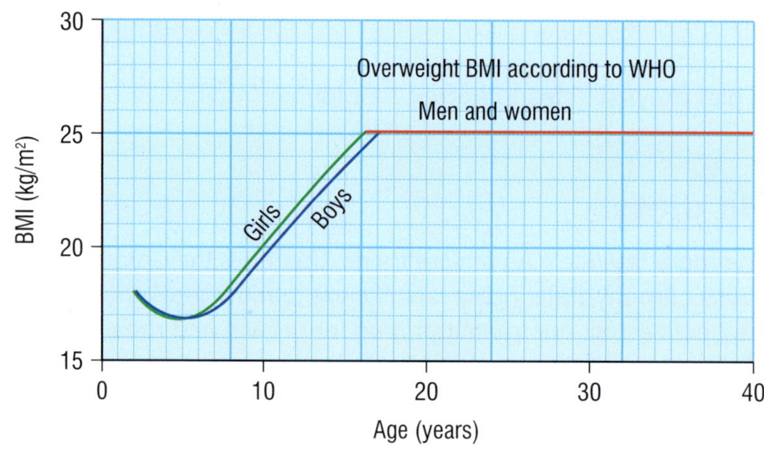

Graph showing BMI information from WHO (World Health Organisation)

$$\textbf{Body mass index (BMI)} = \frac{\textbf{mass (kg)}}{\textbf{height}^2\textbf{ (m}^2\textbf{)}} = \frac{m}{h^2} = \frac{m}{h \times h} \textbf{ (units kg/m}^2\textbf{)}$$

PRACTICAL

Body mass index (BMI)

Weigh yourself in kilograms. Measure your height in metres. Calculate your BMI.

Why doesn't this test apply to pregnant women or muscular athletes?

If your BMI is less than 17.5 or more than 25, that's not good. But you knew that already. Above all, don't worry. It is unhealthy to have anxiety over a less-than-perfect body image. A healthy mental attitude is just as important as physical fitness.

Testing reaction times

A sprinter needs to have a fast **reaction time** to respond to the starter's pistol.

PRACTICAL

Reaction times

Compare your reaction times using an on-line applet, and then using the 'drop-catch' ruler test.

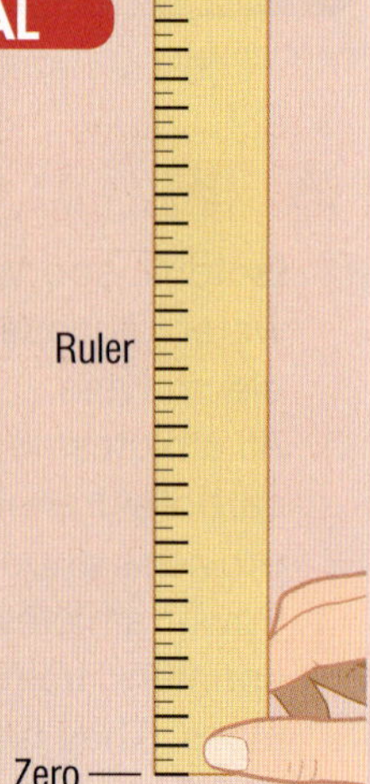

The 'drop-catch' ruler test

Measuring strength

Strength is the ability of a muscle to exert a force for a short time. These involve working against a resistance. There is no single test for strength, as you have different strengths in different muscles.

PRACTICAL

The 'handgrip strength test'

The 'handgrip strength test' involves squeezing a handgrip dynamometer as hard as possible:

1 Hold the dynamometer in line with your forearm and hanging by your thigh. Squeeze without swinging your arm.

2 Repeat after one minute. Then repeat with your other hand.

Testing blood glucose levels using urine

Bayer make a ***glucose-testing strip*** (or ***dip-stick***), called 'Diastix', that changes colour based on glucose concentration.

a) What colour would a glucose-testing strip turn in a solution containing 250 mg of glucose per 100 cm³?

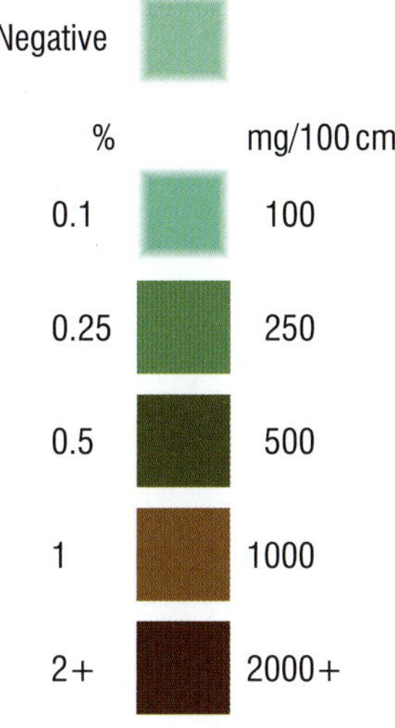

Diastix glucose-testing strip colours

PRACTICAL

Directions for using Diastix

1 Dip the chemical end of the stick into artificial urine and remove immediately.

2 Remove excess urine and wait exactly 30 seconds before comparing with the colour chart.

If you have excessive amounts of glucose in your blood (more than 180 mg of glucose in 100 cm³ of blood), your kidneys filter some of this glucose into your urine. The concentration of glucose in your urine is not what is actually in your blood.

b) Which organs in your body remove excess glucose from your blood?

SUMMARY QUESTIONS

1 Name three sports where good reaction times are important.

2 Why is a low resting heart rate useful to a triathlon competitor?

3 Discuss a sport where endurance, strength, coordination and agility are all important.

4 a) What is more significant to an athlete, the glucose in their blood or their urine?
b) Why do urine glucose levels always lag behind blood glucose levels?
c) Why do urine test strips not show if blood glucose levels are too low?

KEY POINTS

1 You can use base-line measurements when advising on a fitness programme.
2 Body mass index (BMI) is a good indicator of total body fat and reveals your ideal weight, based on your height.
3 The strength of a muscle can be measured using the handgrip strength test.
4 Glucose levels of blood and urine can be measured using a dip-stick method.

5.11 Dressed for success – living in a material world

LEARNING OBJECTIVES

1 Why does sports clothing need to be lightweight, durable and comfortable?
2 Why is friction important in the design of sports equipment?
3 How are aerodynamic shapes and flexibility desirable for sports clothing and equipment?

Drag and friction

Materials scientists are constantly researching, developing and testing new materials to help athletes improve their performance. Speedo, for example, designed the 'fastskin' full-body swimsuit to reduce **friction**. The design, based on the skin of sharks, minimises **drag** in the water.

Air resistance also causes drag. However streamlined body shapes limit friction by reducing the turbulence of the air. Sports scientists often carry out wind tunnel tests to improve the aerodynamics of equipment and the performance of athletes.

Full-body swimsuit

Cyclist training in a wind tunnel

Swimming and track bodysuits are not just a fashion statement. NASA (National Aeronautics and Space Administration) research into blood circulation in space inspired their design. The elastic Lycra in the suit encourages blood to flow more quickly by compressing muscles. Wearing a skin-tight bodysuit also makes you more aerodynamic.

Friction is not always a disadvantage in sports. It provides grip and stops athletes slipping over. However friction does cause wear and produces heat.

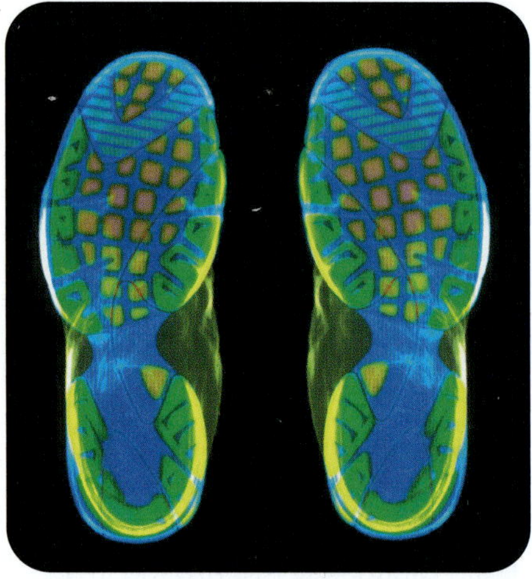
Tread on the soles of some trainers

The textured surface of a basketball

a) Look at the photographs of the trainers and the basketball. How do their surfaces provide grip?

We need different shoes for different events. Sprint shoes need to be light and rigid. Sprinters try to run on their toes (much like many animals). The rigid shoe stops the foot bending that would waste energy. By contrast, distance runners need more comfortable, flexible footwear to absorb muscle vibrations and limit injuries.

Trainer design. Sensors monitor the motion of the ankle, knee and hip. Improvements in the design of the trainer will reduce the chances of injuries.

Fabrics

Cotton is no longer good enough for the modern athlete. Although it acts like a wick or a sponge and draws water away from the skin, new hi-tech fabrics have better wicking properties. Football shirts are a good example. Their yarn, called ***sport wool***, is a mixture of real wool and polyester. The natural fibre is next to the skin to **wick** away the sweat and prevent the player feeling clammy. On the outside, the polyester does not absorb moisture, but is breathable, allowing sweat to pass out of the clothing.

Fibres of a sweat-absorbent fabric 'wick' moisture away to a breathable outer layer.

b) When water from sweat evaporates it cools the body. Why is this? (See page 124.)

At the 1912 Olympics the 100 m was run in 10.6 s. Today the record is under 9.8 s. Runners in 1912 had no spiked shoes, no starting blocks or all-weather polyurethane tracks. Diet and training were little understood.

c) Discuss: World records are broken year on year in athletics. Are today's athletes really any better? Or have they just got better kit?
d) Debate: Is the use of advanced materials in sports equipment unethical?

SUMMARY QUESTIONS

1 a) The model in the photograph on the opposite page is wearing a 'fastskin' swimsuit. Why is a control model also pulled through the water?
b) What would the control model wear?

2 a) How is the cycle helmet in the photograph aerodynamically designed?
b) What is the difference between streamlined and turbulent air?
c) How do cyclists reduce the effects of air resistance?

3 a) Lycra clothes are flexible. How does this help in athletics?
b) How is Lycra an important material in bandages to treat injuries?

4 a) What is wicking and why is it important in sports clothing?
b) Gortex is another wicking fabric that allows sweat to pass out, but it also stops rain entering. Where would Gortex clothes be useful?

5 a) Why is friction important in the design of shoes?
b) Why do running shoes have to be lightweight and durable?

KEY POINTS

1 Friction is important in the design of sports equipment, e.g. grip for the soles of shoes and slip for the aerodynamics of cycle helmets.

2 In your answers think about the advantages of materials and their properties, such as lightweight, durable, comfortable, aerodynamic, waterproof, insulating and wicking (moisture absorbing).

5.12 Same sport, new material

LEARNING OBJECTIVES

1 What are the properties of metals, polymers and ceramics?
2 How are wood, metal, polymers, ceramics and composites used in sports equipment?
3 How do composites benefit from the properties of their component materials?

The LotusSport Pursuit bicycle

Bicycle helmets

The design of the bicycle, with heavy steel frame and equal-sized wheels, is basically the same as 100 years ago. Steel is easily manufactured, inexpensive and strong. However in the 1980s, the use of expensive new materials, such as titanium and carbon fibre, allowed designs to change. A revolutionary aerodynamic design helped Chris Boardman win a 1992 Olympic gold medal. His Lotus bicycle had a strong composite shell. The lightweight frame, made of carbon-fibre in an epoxy resin, was strengthened with titanium at places of high stress.

a) What is a disadvantage of: i) steel? ii) carbon-fibre composite?

From cycling's earliest days there were head injuries. Until the 1970s racing cyclists wore strips of leather padded with wool or sponge.

The first helmet with expanded polystyrene or ***EPS*** (picnic cooler foam) was the Bell Biker.

In 1986 the heavy hard plastic shell was replaced by a Lycra cloth.

To hold the EPS together Mirage reinforced the foam with a nylon mesh.

The next big design step, in 1990, was the reintroduction of a thin shell cover. This time in milk carton plastic or ***PET*** (polyethylene terephthalate), allowing the helmet ***to skid more easily***. Moulding the foam inside a shell also helps ***to hold the foam together*** in an impact.

b) How do both these features improve safety?

The best shape for a helmet in a crash is a ball. However, since Greg LeMond won the *Tour de France* with an aerodynamic design, elongated helmet shapes have become the fashion.

An aerodynamic design

Helmet with air vents

There are four important types of materials used in manufacturing: metals, polymers, ceramics and composites.

Metals

Steel and aluminium, for example, are:

- ***strong*** (needing a large force to break them)
- good thermal **conductors**
- **stiff** (not flexible) and
- ***hard*** (difficult to dent or scratch).

c) Name another metal we use in sports equipment.

Polymers

'Poly-mers' means 'many-bits'; they are organic compounds with long-chain molecules. Wool and cotton are ***natural polymers***. Plastics like polyvinyl chloride (PVC – used for cling-film) and polyester (used for clothing) are **synthetic (man-made) polymers**.

Polymers:

- are flexible
- have a low density
- have a low thermal conductivity.

d) Give the name of another polymer. How could it be useful in a sport?

Ceramics

Ceramics are inorganic, non-metallic materials, e.g. pottery (made of clay) and glass. Ceramics have:

- a high melting point
- low thermal conductivity.

Composites

Composites are made from two or more materials, e.g. metal **alloys** and reinforced concrete. Wood is a natural composite containing cellulose fibres. By combining materials we can use their properties intelligently. For example, glass is strong but brittle. Yet glass fibres in plastic make a strong, tough composite ideal for boat hulls.

DID YOU KNOW?

Carbon-fibre composites are strong but can break on impact. Titanium metal is strong and has a low density. The cockpit of a motor racing car is reinforced by a composite containing a mesh of titanium and carbon fibres.

Golfers can now hit a ball further and more accurately. The shaft of a modern driver is made of carbon-fibre-reinforced composite, which is lighter than a steel shaft. The club head is a hollow shell of titanium alloy. Titanium alloy is stiffer and less dense than steel, so the club head can be larger.

SUMMARY QUESTIONS

1 a) Why is a lighter cycle helmet with air vents better for the rider?
  b) Why are elongated helmets more likely to jerk a rider's neck than rounded helmets?

2 Name a piece of sporting equipment made from:
  a) wood b) steel c) plastic d) clay e) a composite.

3 a) Why does a carbon-fibre shaft help golfers to hit the ball further?
  b) How does a larger club head help with accuracy?
  c) Old wooden club heads absorb moisture. Why is this a problem?

4 a) Classify these as metals, polymers, ceramics or composites:
    i) a brick ii) a nylon rope iii) an aluminium mast iv) a glass-fibre-reinforced plastic hull.
  b) For each type of material: metals, polymers, ceramics and composites, write how to increase performance in a sport.

KEY POINTS

1 Metals are strong, good thermal conductors, stiff and hard.
2 Polymers (natural and synthetic) are flexible, have a low density and low thermal conductivity.
3 Ceramics have a high melting point and low thermal conductivity.
4 Composites benefit from the combined properties of their two or more materials.

5.13 Material of choice

LEARNING OBJECTIVES

1 What natural and synthetic polymers do we use to make sports clothes?
2 What are the pros and cons of natural and synthetic materials?
3 How do the properties of materials affect sports clothing, equipment and footwear?

DID YOU KNOW?

A special material called 'phase change material' can absorb and release heat, keeping the body temperature constant. If the body temperature of an athlete rises after exercise, the gel in the material absorbs energy and melts. If the athlete's temperature falls, the gel refreezes and releases the heat that was absorbed.

Look at your shirt label. What is it made from?

We make clothes using either natural or synthetic polymers. Natural fibres, like cotton and leather, come from living things. Synthetic fibres, like polyester and Lycra, are man-made and often come from crude oil. Often we make clothes from a mixture of natural and synthetic polymers.

Fibre	Advantages	Disadvantages
Natural	• obtained without chemical processing • biodegradable (decays naturally) • wicking (absorbs water, so comfortable)	• limited number available • ignites easily • can shrink in the wash
Synthetic	• inexpensive to make • strong and hard-wearing • can be brightly coloured • dries quickly after washing	• low melting point, melts to skin if burnt and produces toxic fumes • non-biodegradable • can feel clammy and stick to skin

a) Which type of fibre is more resistant to chemical corrosion?
b) How can a synthetic fibre pollute the environment?
c) Why is a synthetic polymer, like nylon, uncomfortable to wear in the summer?

Sports footwear testing using motion sensors

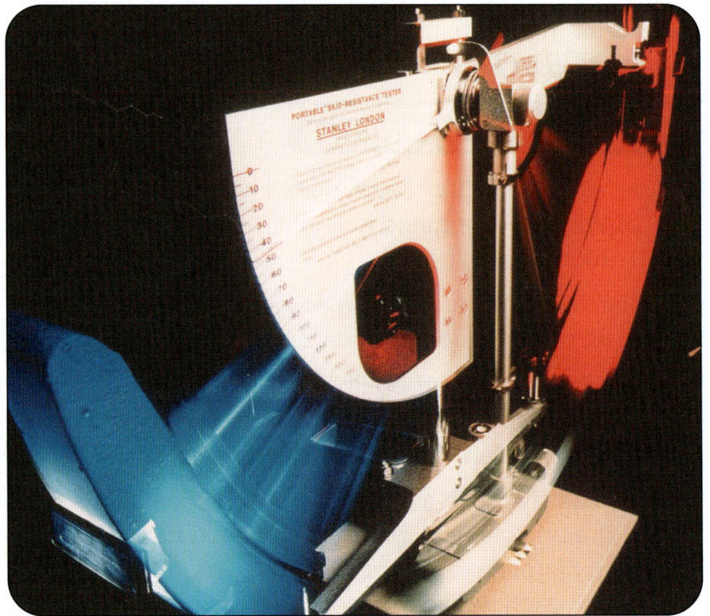

A friction test to measure slipperiness

Tennis rackets

After 700 years the wooden racket was redundant. Wood's lack of stiffness limited power in play.

In 1974 Jimmy Connors won the Wimbledon and US Open finals using an all-steel tennis racket, the Wilson T2000. The key properties of a tennis racket are stiffness, lightweight and vibration dampening. Steel did not dampen vibrations, so added to repetitive stress injuries in muscles and tendons.

In 1980 John McEnroe began four years as world no.1 using the graphite Dunlop Max 200 g. It was made of carbon fibres bound together with a plastic resin. 1998 saw the introduction of the Head Ti6S. This was a mixed composite racket, with strong titanium and light graphite fibres woven together.

Today the choice is usually between:

- a light (250 g) powerful racket with a large 'sweet spot' that feels good
- a heavier (300 g) control racket that encourages flowing strokes and better technique.

Some tennis coaches dislike ultra-light powerful rackets, as they do not encourage players to develop their stroke play and acquire good technique.

d) Why do Head use a titanium–graphite fibre in their tennis rackets?

Head tennis racket design

DID YOU KNOW?

Olympic badminton rules say the bird has to have at least 14 feathers!

SUMMARY QUESTIONS

1 Copy and complete: A natural polymer comes from . . . .. A synthetic polymer is . . . ..

2 a) Give some examples of sports clothes containing Lycra.
b) Why are sports clothes containing Lycra useful?

3 a) Why are wicking materials useful for athletes?
b) What sports clothing needs to be both waterproof and flame resistant?
c) Explain how a mountaineer could benefit from wearing clothes containing 'phase change material'.

4 a) Why is stiffness important in a tennis racket?
b) Why are lightweight rackets more manoeuvrable?

5 How would you use the following apparatus to test the durability or wear-resistance of a fibre?

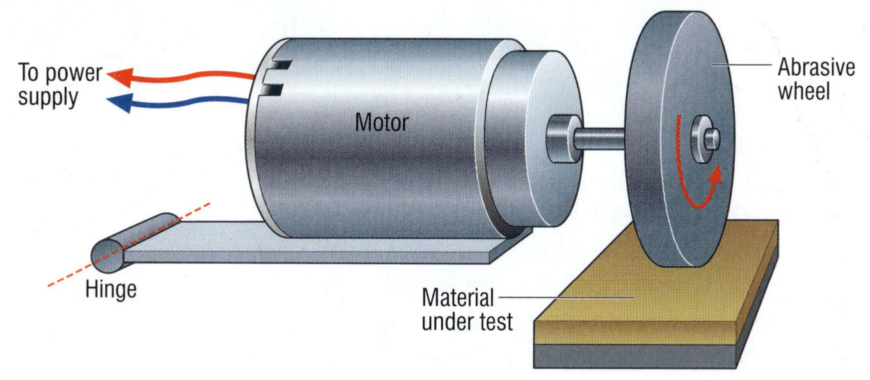

Apparatus to test durability or wear resistance

KEY POINTS

1 Natural fibres, like cotton and leather, come from living things.
2 Synthetic fibres, like polyester and Lycra, are man-made and often come from crude oil.
3 Often we make clothes from a mixture of natural and synthetic polymers.
4 Often we make sports equipment from composites – mixtures of materials.

5.14 Paralympic sports

A Paralympic athlete

Wheelchair design

One person in every 1000 will have a major injury at work this year. One in 40 of these will have a limb amputated. Most of these will be young workers. However these numbers are small compared to the 10 000 children killed and injured by landmines in the world every year.

Many people with disabilities take part in sports. Athletes with a disability use specially adapted equipment to improve their performances.

ACTIVITY

1 Look at the photograph showing a sports wheelchair. Draw out a table showing the reasons for selecting titanium alloy composites to make it.

Property	Advantages of having this property
Lightweight	
Strong	
Stiff	

Landmine victims working in a wheelchair assembly workshop in Cambodia. Expect countries like Cambodia – a country that has had over 40 000 landmine victims – to become the stars of the disabled sports world.

The 2012 London Olympics and Paralympics

In the word 'Paralympics', 'para' stands for 'parallel or equal to', not 'paralysed'. This distinction is important to Paralympic sports stars. The Paralympic Games were first held in 1948 to coincide with the London Olympics. Nurses were employed as referees and doctors handed out the prizes.

Sebastian Coe, for the London Games, said: "29 August 2012 will be a great day in the history of sport. Our vision is to set new standards for services, facilities and opportunities for people with a disability."

ACTIVITIES

2 Discuss the benefits of London hosting both the Olympic Games and the Paralympic Games in 2012.

3 Find out about the 19 sports that take place in the Paralympic Games, and the requirements for Paralympic sports equipment. Design a poster showing how equipment is specially designed for one of these sports.

Prosthetics

Prosthetics is more than just technology. Medical staff must be sympathetic when dealing with their patients. They must also decide whether a ready-made piece of equipment is suitable, or whether their patient needs a purpose-built device. Ill-fitting equipment is more harmful than helpful and, in some cases, is even dangerous. You cannot just give someone an artificial limb and send them on their way. Also children need be seen regularly over a period of years, as they grow.

Testing a prosthetic hand

Prosthetic foot design

A student's water-skiing accident revolutionised prosthetic foot design. The 21-year-old, who lost his leg, teamed up with an aerospace engineer to design a lightweight, strong, yet flexible L-shaped foot. Their prosthetic sprinting 'Flex-Foot', made of carbon-fibre composite, was inspired by the shape of a cheetah's rear leg. Nowadays, 90% of amputee athletes worldwide wear a 'Flex-Foot' in competition. When an athlete puts his or her weight down, energy transfers, which literally puts a spring into their step.

a) What features of the 'Flex-Foot' design help when sprinting?
b) What could happen if the 'Flex-Foot' was stiff as opposed to flexible?
c) Draw out a storyboard that a manufacturer could use to advertise 'Flex-Foot'. Think of an advertising slogan to go with your cartoon.

ACTIVITY

4 **Prostheses** are artificial replacements. How many types of prostheses (man-made replacements) can you think of? Here are two to start you off: an artificial hip, false teeth . . ..

The fastest man on one foot. Marlon Shirley was the first leg-amputee to break the 11 s barrier for 100 m.

ACTIVITY

5 Bend a paper clip into the 'Flex-Foot' shape.
Push down slightly and feel the spring.
Push down more and notice a durability problem.
Explain this problem.

A 'Flex-Foot'

SUMMARY QUESTIONS

1 Complete the sentences using the words below. Words may be used more than once.

aerobic anaerobic biceps blood capillaries contracts debt decrease diaphragm glucagon glucose glycogen heart increase insulin intercostal liver lungs muscles oxygen pancreas pulse relaxes respiration skin sweat triceps

The cardiovascular system refers to the . . . . This together with the . . . . . ., helps . . . . . . and . . . . . . to get to our muscles. Our . . . . . . and breathing rate . . . . . . during exercise. The . . . . . . and . . . . . . muscles in our thorax allow our . . . . . . to ventilate. Respiration may be . . . . . . or . . . . When there is not enough oxygen during exercise, an . . . . in the muscles causes. . . . . to take over. To maintain a constant body temperature during exercise we . . . . . . and . . . . . . in our . . . . . . open up letting more . . . . . . near the surface. We control our blood glucose levels with the help of the hormones . . . . . . and . . . . If our blood glucose levels rise the . . . . . . releases . . . . . ., causing the . . . . . . to convert . . . . . . to . . . . Our . . . . . . and . . . . . . are examples of antagonistic . . . . They work in pairs. When one . . . . . . the other . . . .

2 Complete the sentences using the words below. Words may be used more than once.

basic body carbohydrates diet electrolytes energy exercise glucose glycogen heavy height index isotonic mass muscles pasta protein requirement rice water weight

Our . . . . is 1.3 kcal for every kg of . . . ., every hour. A person's daily energy requirements increase if they are . . . . . . or do more . . . . Athletes often keep a . . . . . . diary. Endurance athletes increase their complex . . . . . . (bread, ..., ...) before competing to increase . . . . . . stores in . . . . . . and the liver. Power athletes have high . . . . . . diets to help build . . . . sports drinks contain . . . . . ., ... and . . . . = ... ÷ . . . . . .^{2} (in units kg/m^2).

3 Complete the sentences using the words below. Words may be used more than once.

aerodynamic areas comfort composites conductivity density durable flexible grip hard high insulation lightweight low man-made melting natural points polymers properties shock strength surface synthetic thermal

Friction provides . . . . shapes are smooth and reduce friction. Sports clothing (and footwear) needs to be . . . . . ., ... and flexible (for . . . . . .). Soles are often . . . . . . absorbing, with a low . . . . . . (for increasing speed). . . . . . . helps maintain body temperature, but large . . . . lose heat more quickly. Cotton and leather are . . . . . . materials. Polyester and Lycra are . . . . . . (or . . . . . .). Metals are . . . . . ., not . . . . . . but stiff, and have . . . . . . thermal . . . . . . and tensile . . . . . . (for support). Ceramics have . . . . are flexible and have low densities and (like ceramics) have low . . . . are made of more than one material, making the most of the . . . . . . of each.

KEY PRACTICAL QUESTIONS

1 Describe how to take base-line measurements of:

a) Body mass index.

b) Heart rate (pulse).

c) Breathing rate.

d) Tidal volume and vital capacity using a spirometer.

e) Glucose content using a dip-stick.

f) Muscle strength using the handgrip strength test.

EXAM-STYLE QUESTIONS

1 a) The simplest way for an athlete to monitor their diet is to keep a diary. Where can athletes obtain information about the foods they eat? (1)

b) Compare the diet of an endurance athlete, like a marathon runner, with a normally balanced diet. (3)

c) i) List the contents of an isotonic sports drink. (3)

ii) Explain how an isotonic sports drink helps a runner in a marathon. (3)

d) Explain one way that the skin of an athlete helps to reduce their core body temperature during a marathon. (3)

2 a) The design of a modern tennis racket enables players to hit the ball with more speed and accuracy than ever before. The frame of a traditional racket was made from natural materials. Traditional rackets had smaller heads and shorter strings. First metal and then graphite replaced natural materials. The new materials are strong enough and light enough to keep longer strings in tension on a larger head. The head of a modern racket is nearly double the size of a traditional racket and the 'sweet spot' is much larger.

i) Give ***one*** advantage of playing tennis using a racket with a larger head. (1)

ii) Name a natural material used to make the frame of a traditional tennis racket. (1)

iii) Name a metal used to make the frame of a modern tennis racket. (1)

iv) Some modern rackets are made of a combination of graphite and glass fibre. What is the name of this type of material? Choose from: ***ceramic, composite, metal, polymer.*** (1)

v) What type of material is used to make the strings of a modern racket? Choose from: ***ceramic, composite, metal, polymer.*** (1)

vi) Some modern rackets are made of a composite material. What is a composite material? (1)

vii) Explain how the design of a modern tennis racket depends on properties of the material used to make the frame of the racket. (3)

b) The performance of a tennis player may also be affected by their choice of footwear. Give ***two*** important features of a tennis player's footwear and explain why each feature is important (2)

c) Explain why tennis players should drink lots of fluid during a match. (3)

3 a) The photograph shows muscles in the upper arm.

i) Label the biceps and triceps. (2)

ii) Outline the changes in these muscles when you bend your elbow. (2)

iii) Explain why muscles like biceps and triceps are called 'antagonistic pairs'. (1)

b) i) During vigorous exercise state what happens to the glucose levels in your blood. (1)

ii) Name the fat that your liver uses to store energy. (1)

iii) Explain how this fat in your liver is converted to glucose. (3)

During vigorous exercise an 'oxygen debt' occurs. Explain the consequences of oxygen debt occurring during exercise. (3)

EXAMINATION STYLE QUESTIONS

1 Look at the picture opposite. Two 'tae kwon do' martial arts athletes wear protective clothing during a fitness test. They breathe into oxygen masks to monitor the 'energy cost' of the exercise.

(a) Explain why they need protective clothing. *(2 marks)*

(b) Give a reason why the helmet:
  (i) can prevent head injuries if it is padded with expanded polystyrene. *(1 mark)*
  (ii) contains air vents. *(1 mark)*

(c) Define what is meant by the phrase 'energy cost'. *(1 mark)*

(d) Give an example of an energy food. *(1 mark)*

(e) Complete the following equation describing aerobic respiration:

glucose + oxygen → *(3 marks)*

(f) State the meaning of the word 'aerobic'. *(1 mark)*

2 (a) When you exercise your muscles need more oxygen. Your breathing and heart rate increase to get more oxygen to your muscles. What organs help to get oxygen to your muscles? *(1 mark)*

(b) Alvin and Bruce decide to monitor their heart rate before, during and after exercise. Results are shown in the graph.

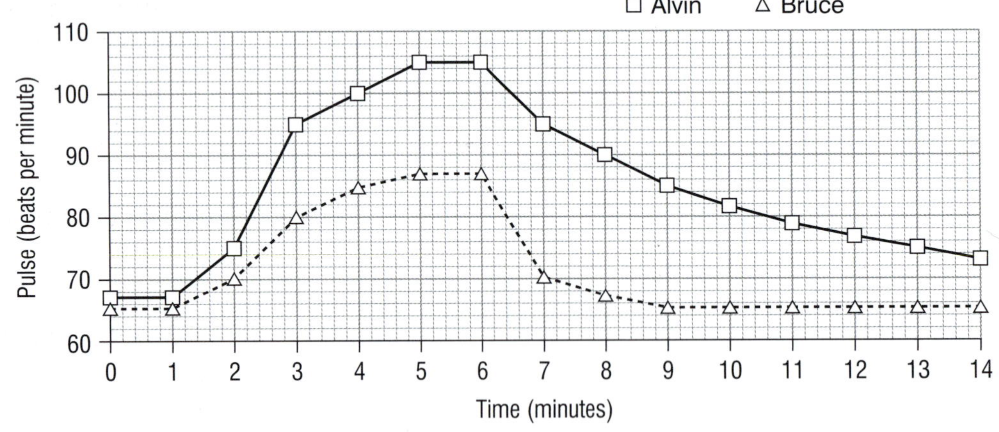

GET IT RIGHT!

– Respiration is ***not*** breathing in and out.

– Respiration is the process of converting glucose to energy.

(i) Who has the higher resting pulse rate? *(1 mark)*
(ii) After how many minutes did they stop exercising? *(1 mark)*
(iii) State three facts from the graph to confirm the fact that Bruce is fitter than Alvin. *(3 marks)*
(iv) What other base-line measurement could Alvin and Bruce have chosen to monitor their fitness? *(1 mark)*

(c) Exercising regularly increases the size of your muscles. Name a muscle that helps you to breathe at a faster rate. *(1 mark)*

(d) Explain why you become fatigued (or tired) during hard exercise. *(2 marks)*

3 (a) (i) Give an example of a synthetic polymer used to make an item of sports clothing. *(1 mark)*
(ii) State the meaning of the word 'synthetic'. *(1 mark)*

(b) Sheila complains that synthetic materials feel clammy when you exercise. She would prefer to wear a garment that is 'wicking'. Explain what she means by the word 'wicking'. *(1 mark)*

(c) Sean agrees with Sheila and adds that if your clothes let you get rid of the sweat from your skin you feel cooler.
(i) Explain how sweating keeps you cooler? *(2 marks)*
(ii) Outline another method your skin uses to keep you cool? *(2 marks)*

(d) Sheila and Sean need more energy during training. Use the following data to calculate Sheila's personal daily energy requirement if she does three hours training in a day:

Sheila's mass = 60 kg
Her basic energy requirement (BER) = 1.3 kcal/hour for every kg of body mass.
For every hour of exercise she needs an extra 8.5 kcal for each kg of body mass.
(3 marks)

GET IT RIGHT!

Examiners often say in a question, 'using the graph' or 'using only the information in the diagram' or 'read the information provided and use it'. In these cases use the information provided. Look at it carefully. Do not panic if you have not met the information before. Simply work logically through the question.

GET IT RIGHT!

During exercise your skin does not go red because you are sweating.

Your skin is redder because capillaries allow blood to flow nearer the surface.

Heat leaves your body because of evaporation (sweating) and radiation (capillary expansion).

GET IT RIGHT!

Do not rush a calculation based on basic energy requirement (BER).

BER = **kg** × 1.3 kcal × **hours**

Substitute, calculate and remember units of kcal at the end.

Remember to bring your calculator to the examination.

5.15 Sports science coursework tasks

LEARNING OBJECTIVES

1 What is involved in my sports science coursework investigation task?
2 How do I maximise my grades?

Health and fitness matter to sports scientists, particularly those parts of the body involved in sport and exercise activities. Sports scientists often study the physiological changes of an athlete during intense training. They use the results to advise the athlete about maintaining personal fitness.

a) Imagine you have skill in a particular sport, yet today you are not successful. Explain why your body may be under-performing.

Sports and materials scientists also develop sports equipment and clothing. They need to choose the best materials for the job.

b) Think about an item of sportswear.
   i) What properties does it need?
   ii) Which materials have the most suitable properties?

Carrying out more than one Unit 3 investigation will help you to perform better in the Unit 2 exam, and progress through the Unit 3 stages of this course. Yet remember, for Unit 3, you only need to produce a report of one practical investigation set in a vocational context. See pages 156–7 for the coursework checklist and mark scheme. 40% of your GCSE marks come from this one report. However you should carry out more than one Unit 3 experiment.

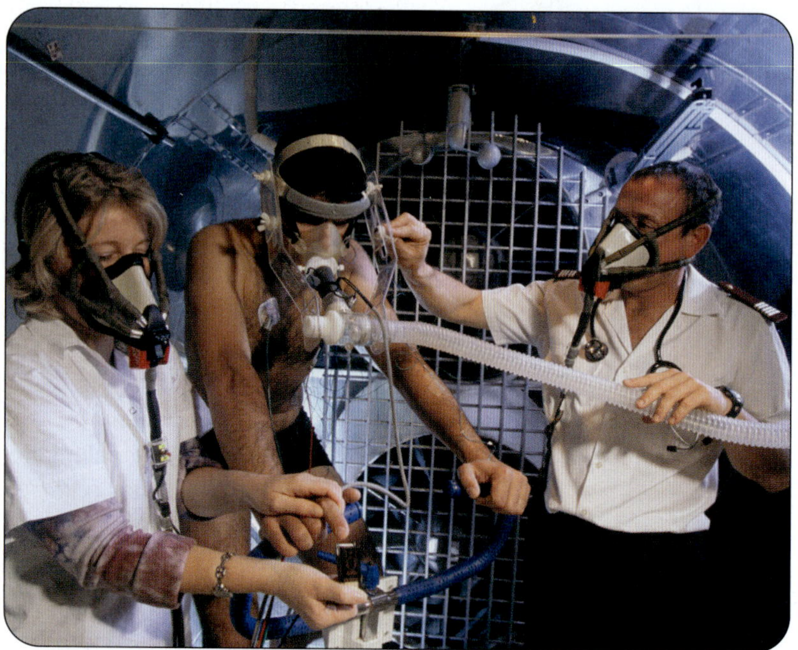

Physiological research. This pressure chamber creates the low pressure and low oxygen levels that occur at an altitude of 8000 m (the height of the tallest mountains on Earth). Low temperature and high wind conditions are also possible. Monitoring equipment records data such as breathing rate, heart rate, gas exchange, blood oxygen content and body temperature.

In the report of your investigation you must:

- Describe the purpose of your investigation.
- Describe how your investigation is connected with a particular sport.
- Include a plan and risk assessment for your investigation.
- Make conclusions from, and evaluate your investigation.
- Explain how a sports scientist might use the results of your investigation.

The sports science investigation options include:

- Either devising, applying, monitoring and evaluating a personal fitness plan for a particular sport or purpose.
- Or investigating the appropriateness of materials that could be used in sport for a particular purpose.

The sports science investigation tasks that we describe in this chapter are:

- Assessing designing a ***personal exercise plan (PEP)*** (pages 148–51). Here you monitor physiological changes in your body during a fitness programme. You may check for and find improvements in your body mass index (BMI), your strength, your power and your endurance. You will analyse factors such as your recovery rate after aerobic exercise, by taking base-line measurements such as heart rate (or pulse), ventilation (breathing) rate, tidal volume and vital capacity. You may also take a fitness test: the ***Harvard Step Test*** or the ***Multi-Stage Fitness Test (MSFT)***.
- Assessing the materials used in trainers (pages 152–5). You will study the grip provided by the sole and the effect of frictional forces in different materials. You will analyse the shock absorbing and moisture absorbing properties of materials. You will investigate how the tensile strength, density, hardness and weathering of materials affect their use.

DID YOU KNOW?

Until 1967, it wasn't illegal for athletes to use performance enhancing drugs.

Fitness testing. The mouth tube leads to sensors that measure oxygen and carbon dioxide. The ***electrocardiograph* (ECG)** monitors her heart.

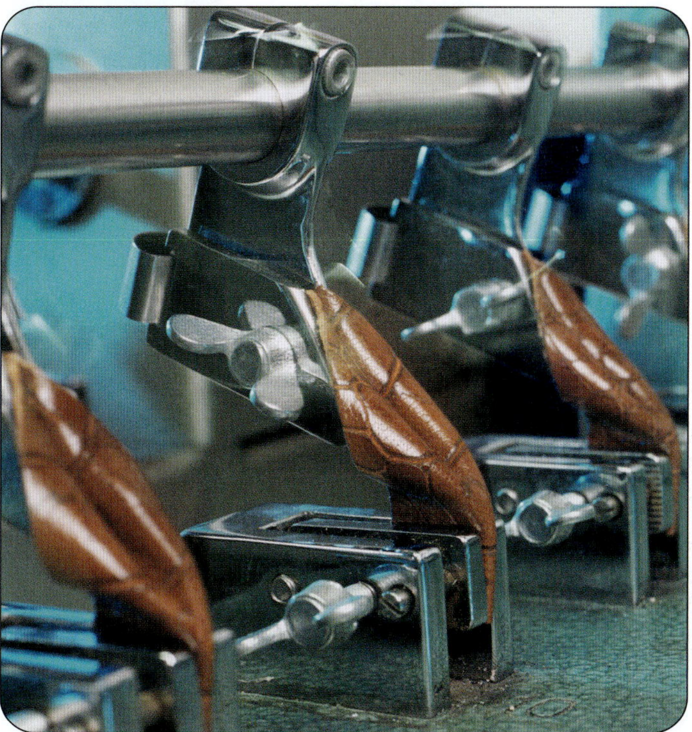

Sports footwear testing at the Adidas research centre. The clamp is twisting and straining samples.

SUMMARY QUESTIONS

1 What are 'physiological changes'? Choose from one of these:

A ***Changes in the response of the organs of our body.***

B ***Changes in the strength of our muscles.***

C ***Changes in our stress levels.***

D ***Changes in our success rates in a particular sport.***

2 Would you prefer to do an investigation based on a ***fitness programme***, or one based on the ***properties of materials***? Explain your reason.

5.16 Designing a PEP: taking base-line measurements

LEARNING OBJECTIVE

1 How do you monitor physiological changes during a fitness programme?

COURSEWORK HINTS

- Design your PEP taking advice from your school or college sports staff.
- Take the same base-line measurements each week during your training.

For your sports science coursework investigation you can choose to design a ***Personal Exercise Plan (PEP)*** for a particular sport or purpose.

Flexibility

Every sport has a different balance of physical needs. The following are seven important factors:

- Strength (like holding or restraining somebody).
- Power (exerting a force in as short a time as possible).
- Agility (like rapid zigzags).
- Balance (controlling body position).
- Flexibility (like doing the splits).
- Endurance (the ability of the heart to deliver oxygenated blood and for muscles to use it).
- Coordination (like good hand-eye responses)

a) Name a sport where each factor is important.
b) Why is endurance relevant to most sports?

There are various standard sports tests that are a measure of strength and power. For example the number of 'sit-ups' in 30 s tests the strength of stomach muscles, while a standing long-jump ('broad jump') tests muscle power. These may be useful additional base-line tests.

Sit-ups strength test

Broad jump power test

In this chapter, you have may looked at the following base-line measurements:

- Heart rate (pulse) page 117.
- Breathing rate, tidal volume (TV), and vital capacity (VC) (page 119).
- Body mass index (BMI), handgrip strength test and reaction times (pages 132–3).
- Sit-ups strength test and broad jump power test, as detailed above.
- Harvard step test and bleep test (pages 150–1).

Safety
Only do these physical activities if you are fit and healthy. Consult your school or college sports staff about safety issues before you complete your risk assessment.

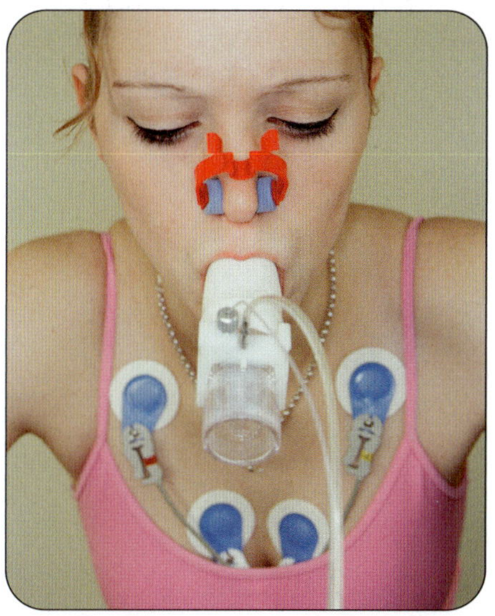

A cardiovascular fitness test. (You will not have complex equipment in your school/college such as electrocardiographs or oxygen consumption meters.)

COURSEWORK TASKS

For your coursework you need to measure the changes that take place in your body over a period of time while you carry out your Personal Exercise Plan. Fellow students, acting as sports scientists, can help you take suitable base-line measurements.

1 Decide which base-line tests are appropriate for your chosen sport or purpose.

2 Carry out the chosen tests. Complete a risk assessment before you start. Remember warming up and cooling down are essential to both training and competition.

3 In your write-up show that you planned your work in well-ordered steps.

4 Complete headings 1–6 of the coursework checklist on pages 156–7 to:

   a) Explain the vocational application of your investigation.
   b) Produce a plan and complete a risk assessment.

5.17 Designing a PEP: aerobic fitness tests

LEARNING OBJECTIVE

1 Which aerobic fitness tests can you use for your coursework investigation in sports science?

For your coursework investigation you need to measure the changes that take place in your body over a period of time while you carry out your PEP (Personal Exercise Plan).

The aim of the following two tests is to monitor your endurance.

PRACTICAL

Harvard Step Test

This test should improve with training.

You need a gym bench, a stopwatch and an assistant.

1 Step up onto a gym bench once every two seconds for five minutes (150 steps). Your assistant helps you keep pace.

2 One minute after finishing the test take your pulse (***pulse 1***) (counting for 30 seconds). Record this in beats per minute (b.p.m.).

3 Two minutes after finishing take your pulse (***pulse 2***) again.

4 Three minutes after finishing take your pulse (***pulse 3***).

5 Calculate your aerobic endurance score using the following formula:

$$\textbf{score} = \frac{\textbf{30000}}{\textbf{(pulse 1 + pulse 2 + pulse 3)}}$$

The sports rating for 16 year-old athletes:

Male	>90 great	80–90 good	65–79 av.	55–64 OK	<55 weak
Female	>86 great	76–86 good	61–75 av.	50–60 OK	<50 weak

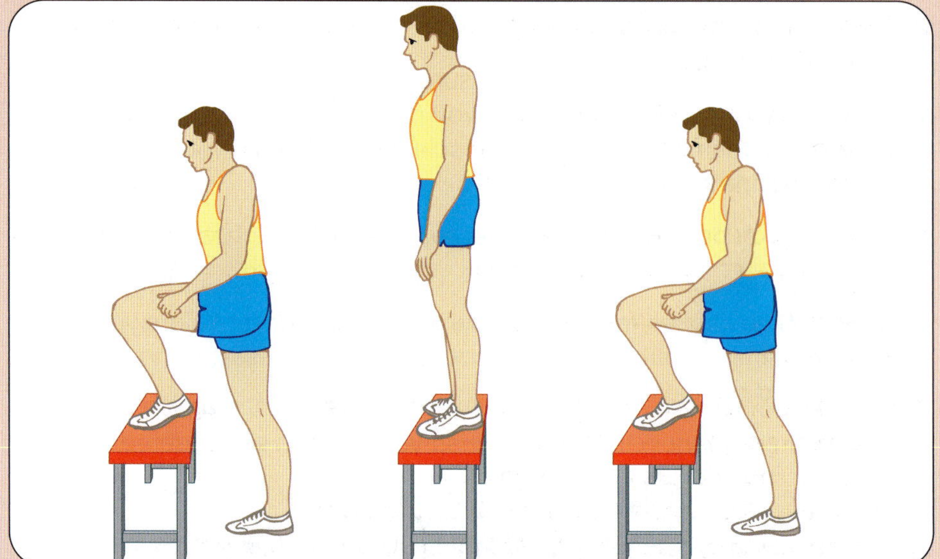

The Harvard Step Test

DID YOU KNOW?

When training ceases, the effect of your training also stops. Your increased strength and mobility will gradually reduce. They decrease at about one third of the rate that you improved during your training.

PRACTICAL

Multi-Stage Fitness Test (MSFT or Bleep Test)

You need a non-slip surface, a tape measure, marking cones 20 m apart, an assistant with the MSFT audio tape/CD, player and recording sheets.

The test involves running between lines 20 m apart in time to bleeps.

At ***Level 1*** your speed is 8.5 km/h. After a minute three bleeps indicates the start of ***Level 2*** at 9 km/h. Your speed must increase by 0.5 km/h to pass each level.

Your score is the level and number of shuttles reached before you are unable to keep up with the bleeps. (As shown in the table.)

Level	Speed (km/h)	Time to run 20 m (s)
1	8.5	8.5
2	9.0	8.0
3	9.5	7.6
4	10.0	7.2
5	10.5	6.9
6	11.0	6.5

Safety: If you are suffering from injury or illness do not take this test. Stop if you feel unwell.

COURSEWORK HINTS

In your PEP you may be advised to use a fitness studio. Your task is to conduct ***pre-test*** and ***post-test*** base-line measurements before exercise and after exercise (as you recover).

a) The Bleep Test is good for games players, who make short turns. Why is it not as suitable for rowers, runners and cyclists?

COURSEWORK TASKS

1 Choose which of the above tests are appropriate for your chosen exercise or sport.

2 Carry out the appropriate tests. Remember to warm up and cool down when taking these tests.

3 Set out suitable results tables to record the data from your chosen base-line tests. Remember to record the dates when you carry out your exercises in your PEP and your base-line tests.

4 What conclusions can you make based on your results? For example, how has your recovery rate improved after aerobic exercise?

5 Explain the importance of base-line tests to assessing the fitness of athletes.

6 Complete headings 8–10 of the coursework checklist on pages 156–7 to:
   a) Select appropriate equipment and carry out your plan, collecting and recording relevant information.
   b) Process the information and make conclusions.
   c) Evaluate your investigation and explain how the findings could be used and applied.

An upper body ergometer

5.18 Investigating materials in trainers (1)

LEARNING OBJECTIVE

1 How do you investigate the materials used in sports trainers?

For your coursework investigation in sports science you can choose to investigate the appropriateness of materials that could be used in sport for a particular purpose.

This section provides you with a basis that you could use for your investigation: How would you investigate the materials used in sports trainers?

The various parts of a pair of trainers are made of different materials, each having different properties. These materials have different functions.

COURSEWORK HINTS

For your Unit 3 coursework you should carry out a range of experiments to assess the properties of materials.

a) Which part(s) of the trainer have to provide each of these?
   i) Grip   ii) Shock absorbency   iii) Moisture absorbency
   iv) Material strength   v) Hardness   vi) Wear resistance
   vii) A low density.
b) What is the advantage of each of these seven factors?
c) What materials could be used to make the trainer's different parts?

Tensile strength is the strength of a material in tension when you try to pull it apart. Tough materials 'neck' or get narrower just before they fracture. Brittle materials crack.

COURSEWORK HINTS

In the tensile (tension test), clamp the line with a Hoffman clip. It acts as a useful hanger for the weights.

PRACTICAL

Tensile (tension) test

1 Add a Hoffman clip to straighten the line (it acts as a useful hanger for the **weights**). Then add a sellotape marker at the zero mark.

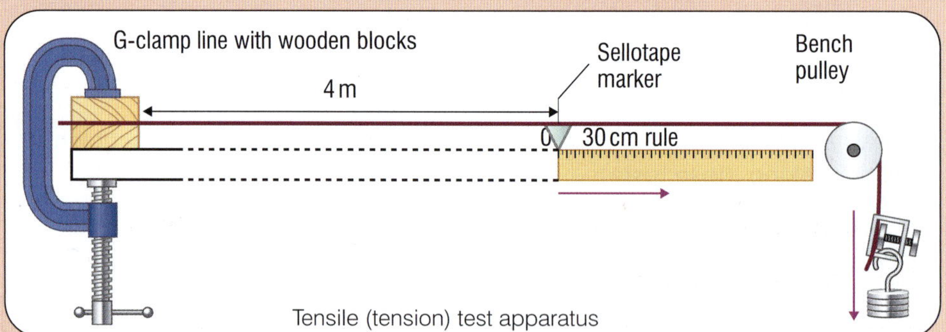

Tensile (tension) test apparatus

2 Compare cotton and polyester thread with solid (mono) and stranded nylon (fishing) line. Note how each stretches as you add loads. Note also the tensile breaking strength (in N). (**Safety**: Avoid masses falling on feet.)

3 Extension:
Measure the radius of each line (in mm) with a micrometer.
Calculate the area of each line (in mm²) using: $A = \pi r^2$.
Calculate the tensile breaking stress of each thread (in N/mm²) using:

$$\textbf{stress} = \frac{\textbf{force (N)}}{\textbf{area (mm}^2\textbf{)}}$$

Friction is useful to provide grip, but is a nuisance as it causes **drag**.

PRACTICAL

Rough / smooth test

1 Secure the material being tested onto the base and arrange the apparatus as shown.

2 Note the force needed for the block to slide. Repeat.

3 Change the material or mass on the block. Repeat.

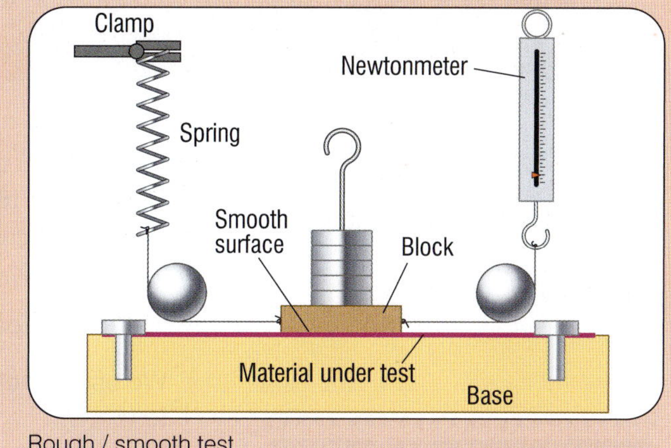

Rough / smooth test

PRACTICAL

Grip test

Arrange the apparatus as shown.

1 Pull the newtonmeter upwards.

2 Note the force needed to just move the trainer. Repeat.

3 Notice the force drops once the trainer moves (because static friction is less than moving friction).

4 Add slotted masses.

5 Repeat.

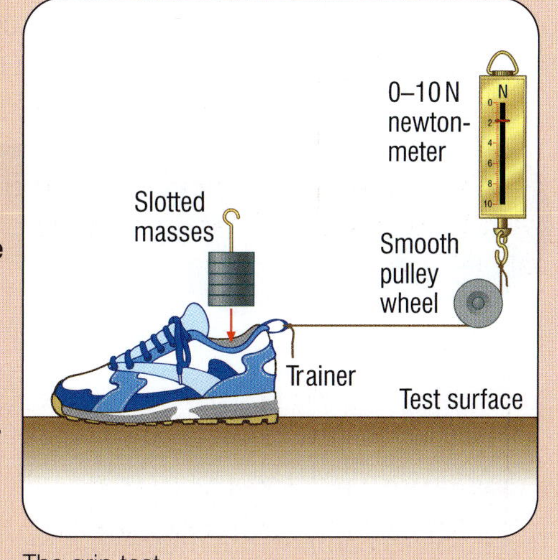

The grip test

PRACTICAL

Wear resistance (or weathering)

1 Set up the apparatus shown below.

2 Time how long it takes for the material to wear through.

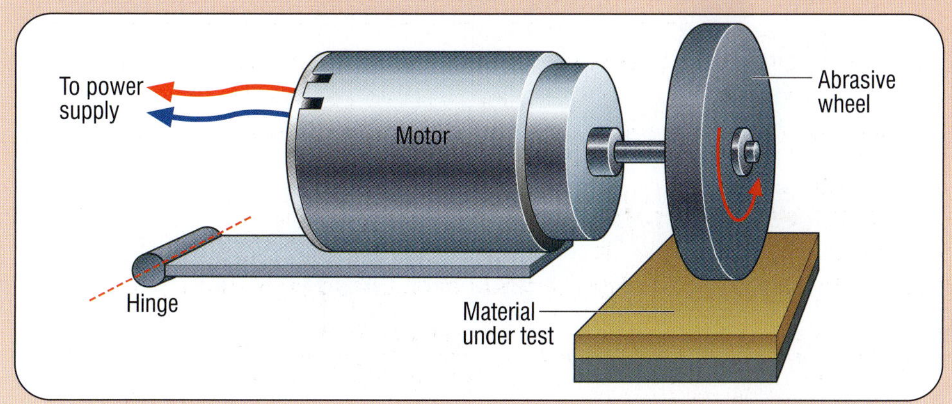

COURSEWORK TASKS

Imagine that you are a sports scientist. You have been asked by a client to assess the materials used in trainers.

1 From the experiments on this and the next spread, plus others based on your research, choose which experiments you would need to carry out.

2 Carry out the relevant tests. Remember that fair tests are important. For each experiment, what will you change and what will you measure?

3 Do a risk assessment before you start each experiment.

4 In your write-up, show your planning in well-ordered steps.

5 What conclusions can you make based on your results?

6 In your evaluation, respond to the brief given by the client.

7 Cover all five skills, A–E (see page 157), in your portfolio.

5.19 Investigating materials in trainers (2)

LEARNING OBJECTIVE

1 How do you investigate the materials used in sports trainers?

For your Unit 3 coursework you are assessing the materials used in trainers. On the pages 152–3 you looked at how to test for tension, roughness/smoothness, grip and weathering. You can also carry out the following experiments for your coursework investigation.

Think about a pair of trainers. Two types of absorbency are important:

- We use ***shock-absorbent materials*** to prevent strain injuries in muscles and tendons.
- We use ***moisture-absorbent materials*** to wick moisture away from sweaty feet.

Look at these three objects below. They have each got the same volume but different thicknesses and surface areas. Just as thicker materials are better for thermal insulation, so thickness affects a material's absorbency.

These objects have the same volume, but different thicknesses and surface areas

DID YOU KNOW?

Penguins huddling in a group stay warm by trapping more air.

PRACTICAL

Shock absorbency test

This involves carrying out a pressure test and an energy test as shown below.

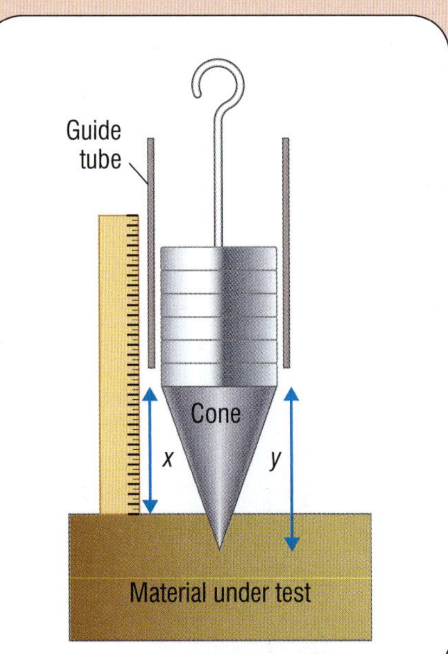

The pressure test: Calculate the depth of penetration from $y - x$.

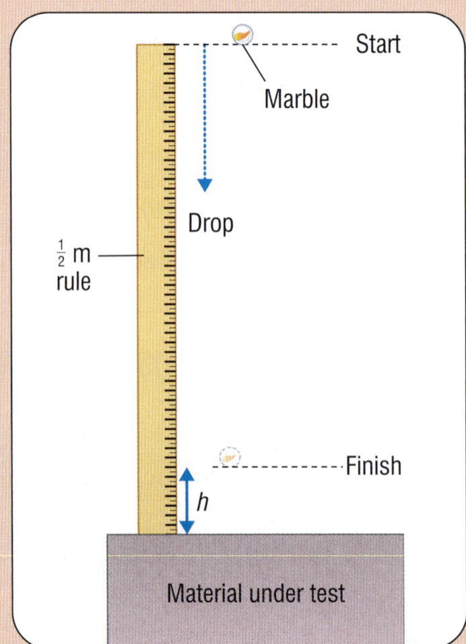

The energy test: Measure the fraction of energy absorbed from height h

COURSEWORK HINTS

Follow the advice in the coursework checklist and mark scheme on pages 156–7.

Remember Unit 3 coursework counts towards 40% of your Additional Applied Science GCSE.

Objects with a ***larger surface area*** have ***more surface molecules*** to absorb moisture. They also dry quicker. You do not scrunch up clothes to dry them! Similarly bodies with a larger surface area cool down quicker.

PRACTICAL

Moisture absorbency

To test for moisture absorbency you can carry out the following two tests.

Soak test

1 Prepare identical-sized squares of different materials.
2 Weigh each one in turn.
3 Immerse them one at a time in 600 ml of water for 30 seconds.
4 Remove them with tweezers and wait until they stop dripping.
5 Reweigh each one and calculate the mass of water absorbed.

Wicking test

1 Secure identical-sized squares of materials onto a test frame.
2 Weigh each in turn.
3 Boil the water. (Take care.)
4 Stand the frame over the boiling water for 3 minutes.
5 Reweigh the material on the frame, calculating the mass of water absorbed.

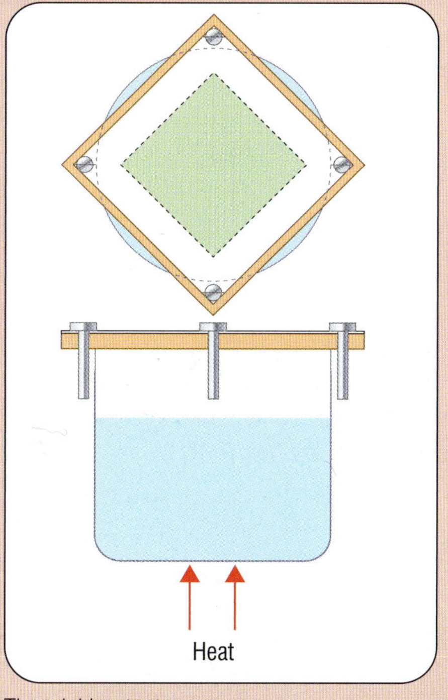

The wicking test

a) What measurements would you take with the equipment in the photograph?

Testing how easily sweat is lost from a trainer

PRACTICAL

Density

1 Measure the length × width × height of blocks of trainer material (in cm³).
2 Weigh the blocks (in g).
3 Calculate their density using:

$$\textbf{density} = \frac{\textbf{mass}}{\textbf{volume}} \text{ (in g/cm}^3\text{)}$$

PRACTICAL

Hardness test

1 Place a ball-bearing on your sample of material. Drop the 1 kg weight onto the ball-bearing carefully.
2 Measure the diameter of the dent with callipers.

Safety: Avoid dropping the falling masses onto fingers and feet.

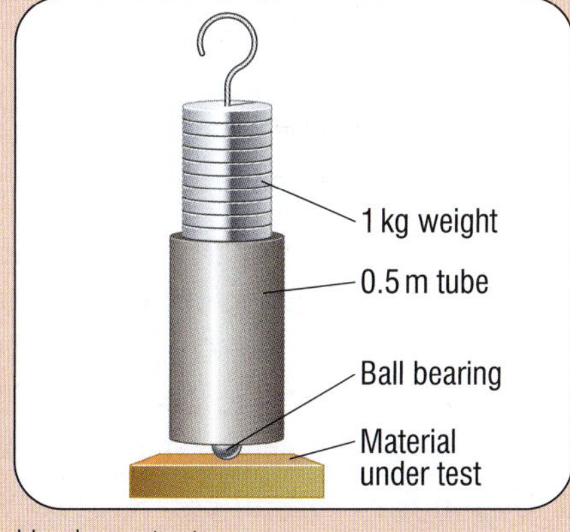

Hardness test

Using scientific skills

For Unit 3 of the AQA Specification you will need to produce a portfolio of evidence, which contains a report of one practical investigation set in a vocational context covering either food science, forensic science or sports science. This accounts for 40% of your Additional Applied Science GCSE mark.

You should carry out more than one investigation as this will help you to perform better in the Unit 2 exam, and progress through the Unit 3 stages of this course. By completing a 'mock' investigation first, your portfolio is more likely to show:

- Increased independence and originality, and greater understanding.
- More ability to plan, research, organise and carry out your experiments, as well as greater ability to record, analyse and evaluate your work.

In the five parts of your portfolio report you should:

A Explain the vocational application of your investigation.
B Produce a plan and complete a risk assessment.
C Select appropriate equipment and carry out your plan, collecting and recording relevant information.
D Process the information and make conclusions.
E Evaluate your investigation and explain how the findings could be used and applied.

Use the following coursework checklist and mark scheme opposite to help you to organise the work for your investigation.

Unit 3 coursework checklist:

1 **Investigation title.**

2 **Aim of the investigation.**

3 **Vocational application:** '*State*' (Stage 1A). '*Describe*' (Stage 2A). '*Research and explain*' (Stage 3A).

So ***research*** first, by using the Internet, or by contacting an organisation directly.

4 **Plan:** '*A series of well-ordered steps*' (Stage 3B).
Anticipate problems that may occur. Include labelled diagrams.

5 **Equipment needed.**

6 **Risk assessment:** See page 17, which shows how to set out a risk assessment.

7 **Results tables:** Remember repeats, and name and units for column headings.

8 **Conclusions:** '*Use the data*' (Stage 1). '*Use and process the data*' i.e. calculations needed (Stage 2). Also use data from research. Write an in-depth reasoned conclusion (Stage 3).

9 **Evaluation:** Refer to the strengths and weaknesses of your work. Consider accuracy, reliability and improvements.

10 **Use in the workplace:** '*Suggest and explain how your findings could be used*' (Stage 3).

The **assessment evidence grid** (mark scheme) for your Unit 3 report follows:

WORKING MORE INDEPENDENTLY		
Stage 1	**Stage 2**	**Stage 3**
You should be able to: • give a simple vocational application of the practical investigation 1A (1 mark)	You should be able to: • describe a vocational application of the practical investigation 2A (2 marks)	You should be able to: • research and explain the vocational significance of the practical investigation 3A (3–4 marks)
• produce a simple plan for the investigation with guidance • carry out a risk assessment for the investigation, given clear guidelines 1B (1–2 marks)	• produce a plan, which with little guidance would enable the investigation to be carried out by another person • carry out a risk assessment for the investigation, given some guidelines 2B (3–4 marks)	• independently produce a plan described in a series of well-ordered steps, which would clearly enable the investigation to be carried out by another person • independently carry out a risk assessment for the investigation 3B (5–6 marks)
• select, with guidance, appropriate equipment for the investigation and use it safely to carry out the plan to collect and record some data/information 1C (1–2 marks)	• select, with little guidance, appropriate equipment for the investigation and use it correctly and safely to carry out the plan to collect and record data/information accurately in a suitable format 2C (4–7 marks)	• independently select appropriate equipment for the investigation and use it correctly and safely to carry out the plan to collect and record data/information accurately and precisely in a suitable format, repeating measurements if necessary 3C (8–12 marks)
• use the data/information collected to make some simple conclusions 1D (1–3 marks)	• use and process the data/information collected to make conclusions 2D (4–6 marks)	• use and accurately process the data/information obtained, and data/information obtained from other sources, to make and present well structured and accurate conclusions 3D (7–10 marks)
• give a simple evaluation of the practical activity 1E (1–2 marks)	• give an evaluation of the practical activity and suggest an improvement to the method • suggest how the findings could be used in the vocational setting 2E (3–4 marks)	• review the work, and present a logical evaluation of its strengths and weaknesses • suggest improvements to the method that would allow the collection of more accurate, precise and reliable evidence • suggest and explain how the findings could be used in a vocational setting 3E (5–8 marks)

- **As a Stage 1 student** you may lack confidence and need considerable guidance when planning, selecting equipment and producing a risk assessment. From basic results, you can make a simple conclusion and evaluation linked to the purpose of your activity. You can state a vocational application of your activity.
- **As a Stage 2 student** you demonstrate some independence but may seek clarification when planning, selecting equipment and producing a risk assessment. You can undertake calculations, interpret data collected and comment on its validity. You can recognise limitations in your methods and can suggest an improvement. You can describe a vocational application of your activity.
- **As a Stage 3 student** you show clearly your independent planning, that anticipates problems that may occur in your investigation. You can independently select appropriate equipment and produce a risk assessment. You appreciate the need to repeat results, complete calculations and rearrange formulas. You can evaluate your work and give clear reasoned conclusions, based on your research and experimenting. You can explain clearly the significance of your findings and how they might be used in the context of food, forensic or sports science.

Glossary

A

Absorption Soaking up. After food has been digested it is absorbed into the bloodstream.

Aerobic respiration The process by which food molecules are broken down using oxygen to release energy for the cells.

Aerodynamic Allowing air or water to flow past smoothly. Aerodynamic streamlining reduces friction.

Alloy A metallic substance formed by combining two or more metals.

Alveoli The tiny air sacs in the lungs which increase the surface area for gaseous exchange.

Anaemia A lack of red blood cells or their haemoglobin, resulting in weariness.

Anaerobic respiration Cellular respiration in the absence of oxygen.

Anorexia nervosa A mental disorder linked to an unrealistic body image and a need for control.

Antagonistic muscles Skeletal muscles that always work in pairs like your biceps and triceps.

Antioxidant A chemical that inhibits oxidation.

Artery A blood vessel that carries blood ***away from*** the heart. (Veins take blood ***to*** the heart.)

Aseptic Without the presence of micro-organisms.

Assessment evidence grid Coursework mark scheme.

Atom The smallest part of an element.

Atria The upper chambers in the heart. (One atrium, two atria.) (There are also two ventricles or lower chambers, the heart being a double pump.)

B

Bacteria Single-celled organisms. Viruses are a thousand times smaller.

Base-line measurements Standard tests to assess fitness of athletes.

Basic energy requirement (BER) For every kilogram of body mass you need 1.3 kilocalories (5.4 kJ) every hour.

Biodegradable Capable of being decomposed naturally.

Biohazard Something that may cause disease in humans.

Biological control Using living organisms (predators of the pests) to control a pest population.

Bladder The sac that fills with urine from the kidneys.

Body mass index (BMI) To assess if people are under- or overweight, the BMI = weight/height^2 (in kg/m^2).

C

Capillaries Thin-walled blood vessels that exchange substances with cells.

Carbohydrate A compound, containing carbon, hydrogen and oxygen, which is a major source of energy in animal diets. Carbohydrates include sugars and starches.

Carbon dioxide A waste product of aerobic respiration, which we breathe out. It turns lime water milky.

Cardiovascular system The heart and its blood vessels.

Cells The building blocks of animals or plants.

Ceramic Inorganic non-metallic materials like pottery (made of clay) and glass.

Chlorophyll The green pigment contained in the chloroplasts.

Cholesterol A substance made in the liver and carried around the body in the blood. High blood cholesterol levels seem to be linked to a high risk of heart disease.

Chromatography A technique used to separate a mixture of substances using a stationary and a moving phase.

Composite Two or more materials mixed together, like glass-reinforced plastic for canoes.

Conduction Heat transfer in a substance due to motion of particles in the substance.

Convection Heat transfer in a gas or liquid due to circulation currents.

Corrosion Wearing away chemically, like rusting.

Corrosive Capable of damaging living tissue, causing burns.

Covalent bonding A way of joining atoms chemically when they share electrons.

D

Deoxygenated blood Blood that is lacking in oxygen, on its way back to the lungs.

Detergent A cleaning agent which makes impurities more soluble.

Diabetes A condition in which it becomes difficult or impossible for your body to control the levels of sugar in your blood.

Diaphragm The sheet of muscle which divides the thorax from the abdomen.

Disinfectant A chemical that kills micro-organisms.

DNA Deoxyribose nucleic acid, the material of inheritance.

Drag force A force opposing the motion of an object due to fluid (e.g. air) flowing past the object as it moves.

E

E-number A number given to a food additive to identify it.

Electrons Negative particles that orbit the nucleus of an atom.

Element A substance made up of only one type of atom.

Emulsifier A substance which stops the two liquids in an emulsion from separating.

Environment Our surroundings, which affect animal and plant life. Pollution threatens the environment.

Eutrophication The process by which water sources (e.g. lakes) become enriched with nutrients. It eventually causes the death of organisms that live in that habitat.

Evaporation Energetic liquid molecules turning to a gas.

Explosive Likely to blow up.

F

Fat An oily compound that provides a store of energy. It also protects organs and insulates against heat loss.

Fats (saturated and unsaturated) See **saturated fats** and **unsaturated fats**.
Fermentation Another name for anaerobic respiration of yeast (see also **anaerobic respiration**).
Fertiliser A substance which is added to the soil to aid plant growth.
Fibre A food group which is not digested by the body, but is important to help remove waste products from the body.
Filter Filter paper separates an insoluble solid from a liquid. The kidneys filter the blood to make urine.
Flammable Capable of burning easily.
Flavour enhancer A food additive that is used to bring out the flavour in a wide range of foods, without adding a flavour of its own.
Flavourings These are added to foods, usually in very small amounts, to add a particular taste or smell to a food.
Food additive A substance added to food to improve its flavour, texture or shelf-life.
Food colouring A chemical that can be added to foods to enhance a particular colour.
Food poisoning An infection caused by food containing harmful bacteria.
Food Standards Agency The independent food safety watchdog set up to protect the public's health and consumer interests, to promote good eating habits, and ensure that our food is safe and correctly labelled.
Friction force A force opposing the relative motion of two surfaces when they are in contact with each other.
Fungi Organisms such as moulds, yeast and mushrooms. Fungi lack chlorophyll.
Fungicide A chemical used to kill fungi.

G

Genetically modified (GM) Plants and animals that have had their genes altered to modify their characteristics.
Glugacon Hormone involved in the control of blood sugar levels.
Glycogen Carbohydrate store in animals, including the muscles, liver and brain of the human body.

H

Haemoglobin The red pigment which carries oxygen around the body.
Herbicide A chemical used to kill weeds or other plants with leaves.
Hydroponics Growing plants in water enriched by mineral ions rather than soil.

I

Immune system The body system which recognises and destroys foreign tissue such as invading pathogens.
Ingredients The items that are used to make a food.
Insulin Hormone involved in the control of blood sugar levels.
Intensive farming A method of farming to minimise costs yet maximise production.
Iodine number The mass of iodine in grams which is consumed by a chemical substance. The iodine number is used to test for the presence of unsaturated fats in foods.
Ion A charged atom.
Irritant A substance that causes inflammation (reddening) to skin, eyes, etc.
Isotonic (isoelectronic) sports drink A drink containing water, glucose and dissolved Na^+ and K^+ salts.

J

J (joule) A unit of energy. (1 kcal = 4.2 kJ.)

M

M A way of measuring the concentration of solutions in moles per litre or mol/dm^3.
Metabolism All the chemical processes that occur in your body.
Microorganisms Bacteria, viruses and other organisms which can only be seen using a microscope.
Minerals Essential elements required in small amounts for healthy animal or plant growth.
Molecule A particle with two or more atoms.

N

NDNAD National DNA Database.
Nutrients A substance that provides nourishment.
Nutrition The process of absorbing useful chemicals from raw materials.
Nutritionist A dietician who studies how food nourishes people.

O

Organ A group of different tissues working together to carry out a particular function.
Organic A substance which contains (mainly) carbon in combination with other elements.
Organic farming Production of food and other materials without the use of chemicals. This method provides animals with an agreed standard of living.
Organism A living individual, such as a plant, animal or bacterium.
Oxidising agent A substance which helps other substances to burn or explode.
Oxygen debt This is the amount of oxygen required to break down lactic acid produced by anaerobic respiration.

P

Pathogens Micro-organisms which causes disease.
Pesticide A chemical used to kill pests.
pH A scale running from 0 to 14 that describes the degree of acidity of a solution.
Physiologist, sports A scientist who studies the functioning of athletes' bodies.
Polymer A substance consisting of very large molecules made of smaller identical molecules called monomers.
Portfolio Your sample of work (your coursework file – marked by your teacher and sent to an external moderator).
Precipitate A solid material produced from a solution.
Precipitation See **precipitate**.
Preservatives Chemicals used to help keep foods safe for longer.

Protein Food group which provides the body with material for tissue repair, growth and energy. Proteins are complex organic compounds made from amino acids, and form the basis of living tissue.

Q

Qualitative analysis Any way of detecting what substances are made of that does not involve calculations.

Quantitative analysis An analysis involving calculations, like a titration.

R

Radiation Energy carried by waves.

Reaction time The time it takes you to respond to a stimulus.

Refractive index A measure of light-bendability. Refractive index = $\sin i / \sin r$.

Respiration The process by which food molecules are broken down to release energy for the cells.

Risk assessment A health and safety check completed before an activity, stressing hazards (dangers), risks (probabilities of harm and seriousness of consequences), control measures (safety precautions) and emergency action (first aid).

S

Saturated fats Fats derived from animal sources. Eating too much saturated fat can raise cholesterol levels leading to heart disease.

SOCO (and CSI) Some police authorities call 'scene of crime officers' by the name 'crime scene investigators'.

Sodium chloride (salt) A widely used flavour enhancer, which can raise blood pressure if taken in too large an amount.

Solvent A liquid in which some solids will dissolve.

Starch A source of carbohydrates obtained from cereals (like wheat and rice) and potatoes.

Sterile Free from living organisms.

Sterilisation Preventing reproduction, or the process of making apparatus sterile.

Stiff A stiff material is not flexible but stays rigid.

Sugars A subgroup of carbohydrates. They dissolve in water and have a characteristic sweet taste.

Sweeteners Chemical additives which are added to foods to provide sweetness without the addition of sugar.

Synthetic Man-made, rather than naturally occurring.

T

Tension A force that resists pulling something apart (rather than squashing or compressing it).

Thickeners Substances which are used to thicken foods.

Thorax The upper (chest) region of the body. In humans it includes the ribcage, heart and lungs.

Tidal volume (TV) The volume of air you breathe in and out normally.

Titration A method for measuring the amount of substance in a solution.

Toxic Poisonous, i.e. causing death or damage to health.

Trace evidence Evidence that exist in minute amounts, such as hairs, paint flecks, pollen, seeds, dust, soil, fibres and tiny flakes of glass.

U

Unsaturated fats Fats derived from plants, which are considered to be healthier than saturated fats.

V

Ventilation Breathing in (inhaling) and out (exhaling).

Vital capacity (VC) The maximum volume of air you can force out of your lungs, after breathing in as hard as you can.

Vitamin deficiency Lacking in a specific vitamin.

Vitamins Compounds needed in small amounts for normal growth of the body.

W

Weight The force of gravity on an object (in newtons).

Wicking Water absorbing property which can be used for sports materials.

Y

Yeast Single-celled fungi that can cause the fermentation of carbohydrates to produce carbon dioxide and ethanol (alcohol).

Index

Acknowledgements

Alamy/Buzz Pictures 113br, /**Chris George** 130, /**Craig Cozart** 13, /**Foodfolio** 32b, 39t, /**Forrest Smyth** 129, /**Fotovisage** 23, /**Holt Studios International Ltd** 31b, /**Ian Miles-Flashpoint Pictures** 78br, /**Janine Wiedel** 20, /**Justine Kase** 6r, /**Mark Harwood** 78bl, /**Mikael Karlsson** 79b, /**Patrick Ward** 108, /**Photofusion** 32l, /**Photolibrary Wales** 6l, /**Pictorial Press** 94c, /**Popperfoto** 8, /**Profmedia, CZ.sco** 38l, /**Royalty Free** 12l, 31t, 38b, 60, /**Steven Shepherd** 69tl, /**Steven Shepherd** 78tl, /**Steven Shepherd** 97, /**Vehbi Koca** 112b, /**David Reed** 136bl; **Chris Honeywell** 18bl; **Corbis/Andrew Brookes** 69tr, /**Herbert Spichfinger** 113bl, /**Hulton Archive** 9r, /**Mike King** 9b, /**Reuters** 131t; **Corbis: V94 (NT)** 18t; **Corel 364 (NT)** 131b; **Corel 541 (NT)** 2bl; **Digital Vision 1 (NT)** 16; **Digital Vision 6 (NT)** 14c; **Empics** 136tl; **Frank Lane Picture Agency** 66br, bl, c; **Gerry Blake** 9tr, c, cr, 22tr, bl; **Getty Images/AFP** 141l, /**Iconic** 136br, /**Photonica** 139, /**Reportage** 134tl, /**Sport** 141b, /**Stone** 134tr; **Ingram ILR V1 CD3F (NT)** 41; **Martyn Chillmaid** 64c; **NHPA** 66tl; **Photodisc 28 (NT)** 112t; **Photodisc 38A (NT)** 2tc; **Photodisc 72 (NT)** 94b; **Photodisc 72 (NT)** 2tl; **Rubberball WWW (NT)** 2tr, br; **Science Photo Library** 46t, 55, 61t, 79t, /**A C Seinet** 49b, /**A J Photo** 11br, /**A J Photo/Hop Americain** 88b, /**Adam Gault** 59br, /**Alex Bartel** 22tl, /**Alexander Tsiaras** 14br, /**Alexis Rosenfield** 2l, /**Alfred Pasieka** 72bl, 80b, 90l, /**Andrew Lambert** 37r, 62(all), 84l, cl, r, 86l, r, 104t, /**Andrew McClenaghan** 43, 61b, /**Andrew Syred** 27tl, 77tr, /**Art Wolfe** 154, /**Biophoto Associates** 35r, t, /**Brian Bell** 28br, 113 tl, /**Bryan Peterson/Agstock USA** 51b, /**BSIP, Raguet** 14, /**BSIP, Keene** 140tl, /**BSIP, LECA** 115, /**BSIP, Roux** 30b, /**BSIP, Chassenet** 143, /**BSIP, Mendel** 27bl, /**Chris Sattlberger** 114, /**Christian Darking** 126b, /**Cordelia Molloy** 18tl, 40b, /**Cristina Pedrazzini** 40tl, /**Crown Copyright/Health & Safety Laboratory** 138br, /**David Aubrey** 48, /**David Constantine** 140b, /**David Parker** 12t, /**David Scharf** 44b, /**David Taylor** 84cr, /**Dr Gary Gaugler** 46b, /**Dr Jeremy Burgess** 76 bc, /**Dr Jurgen Scriba** 87, 93, /**Dr Kari Lounatmaa** 5, /**Dr Tim Evans** 1, /**Edward Kinsman** 125b, /**Erika Craddock** 126t, /**Geoff Tompkinson** 2bc, 29t, 75, /**Geogre Lepp/Agstock** 51t, /**Gusto** 30l, 102 b, /**Hank Morgan** 14cl, /**Holt Studios International Ltd** 49t, /**Hugh Turvey** 134bl, /**Hybrid Medical Animations** 116, /**Ian Boddy** 122, /**James Holmes/Thomson Laboratories** 82t, /**Jim Varney** 104b, /**John Greim** 151, /**Laguna Design** 80l, /**Martin Bond** 11tl, /**Martyn F Chillmaid** 27tr, /**Mauro Fermariello** 27br, 28bl, 69bl, 70r, 71b, 72tr, r, 73cr, bl, 76cl, bl, 77b, 88l, 90b, 92b, 110, 128, 144, /**Maximilian Stock Ltd** 26, 58tl, 67, /**Melau Kulyk** 69cr, /**Michael Donne** 39c, 68, 71t, 73t, 100, 101, 107, /**Michael Jones** 53r, /**Michael Viard, Peter Arnold Inc** 82l, /**Michael Viard, Peter Arnold Inc** 92t, /**National Library of Medicine** 14tr, /**Oscar Burriel** 148, /**Pascal Geotgheluck** 76tr, /**Paul Rapson** 58bl, /**Peter Menzel** 29b, 47r, 53t, /**Phillipe Psaila** 63, 135, 138bl, 146, 147r, 155, /**Professor P M Motta, G Macchiarelli, S Anottola** 37t, /**Robert Brook** 52tl, /**Rosenfield Images Ltd** 44l, 45t, /**S J Krasemann** 125t, /**Samuel Ashfield** 113tr, 121, 147l, 149, /**Scimat** 45c, /**Simon Fraser** 52b, /**Simon Lewis** 11tr, /**Sotiris Zaferis** 64b, /**TEK Image** 64t, 69cl, 74, 89b, t, /**Tony McConnell** 134br, /**Veronique Leplat** 102t, /**Volker Steger** 47t, /**Volker Steger, Peter Arnold Inc** 76br, /**James King-Holmes** 73br, 141t; **Steve Diamond** 19; **Topfoto** 91; **Tyco Fine and Integrated Solutions** 18b.

Additional picture research by Alison Prior.

Thanks are also due to:

Stewart Chenery; Laura Halstead, Lotus Engineering, Group Lotus plc; Mike Powell-Evans, Adnams plc; Jim Burzio (Head of Scientific Services) and David Stag (Senior SOCO), Suffolk Police